⟩ The New Labrador Papers of Captain George Cartwright ⟨

Captain George Cartwright (1739–1819). Colour oil by W. Hilton, ca. 1791. (Private collection, reproduced with permission. Damage to canvas over dog's head and some discolouration of original have been altered).

The New Labrador Papers of Captain George Cartwright

INTRODUCED AND EDITED BY
MARIANNE P. STOPP

McGill-Queen's University Press
Montreal & Kingston · London · Ithaca

© McGill-Queen's University Press 2008
Introduction and scholarly apparatus © Marianne Stopp 2008
ISBN 978-0-7735-3382-0

Legal deposit second quarter 2008
Bibliothèque nationale du Québec

This book has been published with the help of a grant from the Canadian Federation for the Humanities and Social Sciences, through the Aid to Scholarly Publications Programme, using funds provided by the Social Sciences and Humanities Research Council of Canada.

McGill-Queen's University Press acknowledges the support of the Canada Council for the Arts for our publishing program. We also acknowledge the financial support of the Government of Canada through the Book Publishing Industry Development Program (BPIDP) for our publishing activities.

LIBRARY AND ARCHIVES CANADA CATALOGUING IN PUBLICATION

Cartwright, George, 1739–1819.
The new Labrador papers of Captain George Cartwright / edited by Marianne P. Stopp.

Includes bibliographical references and index.
ISBN 978-0-7735-3382-0

1. Cartwright, George, 1739–1819. 2. Frontier and pioneer life – Newfoundland and Labrador – Labrador. 3. Building – Newfoundland and Labrador – Labrador – History – 18th century. 4. Material culture – Newfoundland and Labrador – Labrador – History – 18th century. 5. Trapping – Newfoundland and Labrador – Labrador – History – 18th century. 6. Salmon fishing – Newfoundland and Labrador – Labrador – History – 18th century. 7. Sealing – Newfoundland and Labrador – Labrador – History – 18th century. 8. Inuit – Commerce – Newfoundland and Labrador – Labrador – History – 18th century. 9. Indians of North America – Commerce – Newfoundland and Labrador – Labrador – History – 18th century. 10. Labrador (N.L.) – Description and travel. 11. Labrador (N.L.) – History – 18th century. 12. Merchants – Newfoundland and Labrador – Labrador – Biography. I. Stopp, Marianne P. II. Title.

FC2193.3.C37A3 2008   971.8'202092   C2007-907204-6

Set in 10.5/14 Filosofia
Book design and typesetting by Garet Markvoort, zijn digital

❧ CONTENTS ❧

ACKNOWLEDGMENTS

Foremost I wish to thank John Cartwright, Johannesburg, South Africa, for generously granting permission to publish his ancestor's papers and for providing quality copies of the portraits of George Cartwright and the Inuit woman and other items.

My husband, Richard, and my son, Thomas, are of course lovingly implicated, and I gratefully acknowledge their support throughout.

Ingeborg Marshall has ensured that yet another valuable part of Canadian history was both found and brought to publication. She has my deepest gratitude for entrusting me with this work and for help with all aspects of its completion, which included providing research material collected over the years at the British Admiralty Office, the Dorset County Records Office, and other institutions, for lists of references that she felt were important and needed checking, for establishing the link with the Cartwright family, and for helping with many minor items such as eighteenth-century notations and currency. Her friendship over the years has been a cornerstone.

For their interest and efforts on behalf of this research, thanks are extended to, alphabetically: Eduardo Alvarez for help with the images; Oriana Barkham and Selma Barkham for transcription of a piece of correspondence; Marc Cockburn, Library and Archives Canada, for help with images; Anne Duggan for editorial skills; Patricia Kennedy, Library and Archives Canada, for her interest in Cartwright; Eva Luther for her dedication to the history of her Labrador homeland; Andrew Nicholson, Nottinghamshire History Society, for confirming Cartwright's burial date (although his precise date of death remains in question); Ken Reynolds, Provincial Archaeology Office, Newfoundland, for generously sharing research references over the years; Barbara Rieti, for offering solid stouters and shores throughout; Joan Ritcey, Centre for Newfoundland Studies, Memorial University, for facilitating access to the Cartwright material; Leah Sander, Hudson's Bay Company Archives, Winnipeg, Manitoba, for finding William Paulson. Also, thanks to Meryl Oliver and Judith Romard, Parks Canada; Fred Schwarz, Black Spruce Consulting, for copying source material; Mary Scott, for ensuring that my metaphorical house is well framed and "chinced"; Marie Wadden for many encouragements; and finally, Leo Zrudlo, architect, and Mark Glassford and Susann Myers, architects with Public Works and Government Services Canada, whose collective interest in early North American construction helped tremendously in making sense of "walplates," sills, and lining boards. Through a simple sketch, Susann Myers interpreted Cartwright's building instructions into a language that I could understand.

It is in the nature of old documents that they lie forgotten for many years, eventually to be lost altogether. Yet I am certain that there is more Cartwright material to be found. There is indeed still much to be written about Labrador. McGill-Queen's University Press is fully acknowledged for its dedication to presenting books on the history of a fascinating province, which was my home for twenty-five years. This volume could not have come about without the inspired and expert input of Jonathan Crago, Maureen Garvie, Garet Markvoort, and Joan McGilvray and the support of the Aid to Scholarly Publications Programme.

Fig. 1   "George Cartwright Visiting His Fox Traps."
Engraving by T. Medland based on the oil portrait by
W. Hilton, from Lysaght (1971, figure 22).

shown in figure 14. In the early 1980s I transcribed the Cartwright papers
for Dr Marshall, and a decade later received permission to publish them.
A fuller assessment, however, had to undergo a lengthy period of gestation
on my part and owes its completion to Dr Marshall's continued interest in
having this material published.

In the text to follow, the terms "the coast" and "the south-central coast"
are interchangeably used to refer to the area where George Cartwright
lived, namely the coastal section south of Hamilton Inlet between Chateau Bay and Sandwich Bay, Labrador. Adjoining areas referred to in the
text are the Strait of Belle Isle, also called the Straits region, that part of
Labrador on the western side of the waterway. The North Shore, the Côte

The Cartwright papers were discovered by Ingeborg Marshall in 1979. Author of *A History and Ethnography of the Beothuk* (1996), Dr Marshall was at that time researching the extinct Beothuk Indians of Newfoundland and attempting to find related documents in public and family archives. After considerable searching, she eventually located the Cartwright family archive in South Africa, which held material relating to Major John Cartwright who had compiled some of the earliest maps and observations of Beothuk settlements in interior Newfoundland. Through the encouragements of Dr Leslie Harris, then vice-president of Memorial University, Dr Marshall was able to travel to South Africa to view the collection, which turned out to be unexpectedly valuable and relevant to Canadian history. It held a variety of material relating to the Cartwrights of Nottinghamshire, England, who counted among their members Captain George Cartwright (1738–1819) of Labrador fame; his brother Major John Cartwright (1740– 1824), a British naval officer, one of the first Europeans to describe aspects of the Beothuk of Newfoundland, and later a social and political reformist; and a third brother, the Rev. Edmund Cartwright (1743–1822), a clergyman who became famous as the inventor of the power-driven loom. With the generous permission of John Cartwright and his mother, Mrs Patricia Cartwright, Dr Marshall was able to have microfilm copies made of the Labrador papers, to photograph certain items, and to prepare an inventory of this extensive collection (Marshall 1979).[1] The papers presented here are based on the microfilm copy and the original material in the Cartwright family archive.

Ingeborg Marshall's discovery included an oil portrait of George Cartwright by W. Hilton, published here as the frontispiece, which was the template for the well-known black and white lithograph by T. Medland (figure 1). She also discovered the small watercolour of an Inuit woman,

When faint & weak with walking & long fasting, break a cake of bread & put it into water, whilst it is soaking, scrape your tongue & wash your mouth as above directed, then eat the bread & drink one mouthful of water only. On arriving at home, never indulge in taking a drink or cold liquor within an hour, but take the chill off it with a hot poker, or a mixture of what is warm. Never eat much, or one bit of flesh-meat, until you have finished your Days journey, for it will oppress you and cause unquenchable thirst. Spruce beer is the best common drink, and an excellent anti-scorbutic.

When you get into great Danger, deliberate only on the best means of extricating yourself & stick to that without changing, unless circumstances alter. The Man who looks Danger baldly in the face will escape it nine times in ten but the undetermined seldom will. He who wants presence of Mind & resolution in danger shows the strongest marks of cowardice, and increases his danger very considerably. Foolhardiness is no proof of courage but a very great one of want of sense.

In short, observe everything, and carefully enter all your observations in your Diary.

Written for someone we know only as "W," who was clearly headed for the New World fur trade, the above words are both Captain George Cartwright's advice and his creed. "Memorandum for W," fully transcribed later in this volume, tells us as much about the personal code of the man who wrote it as it does about the demands and challenges of frontier life. Cartwright is revealed as an exacting and fastidious individual, ultimately courageous but also unquestionably idiosyncratic.

George Cartwright is a familiar icon of the settler landscape in southern Labrador, but he remains a relatively obscure Canadian historical figure. As one of the earliest British merchants in Labrador, he operated a series of sealing, salmon, and cod-fishing rooms, or posts, between Cape Charles and Sandwich Bay between 1770 and 1786. He would be less known to us had he not followed his own dictum to observe and record everything. For nearly sixteen years he kept an account of his life in Labrador, which he published in 1792 as *A Journal of Transactions and Events during a Residence of Nearly Sixteen Years on the Coast of Labrador* (hereafter referred to as the *Journal*). The eminent Newfoundland historian D.W. Prowse aptly described this three-volume work as both ponderous and "one of the most remarkable books ever written" (Prowse [1895] 2002, 599).

CHAPTER ONE

The Labrador Writings of George Cartwright

Learn the Eskimeau language.

Provide a Quadrant & an Ephemeris to take Latitudes & Longitudes, and take Daily observations during the passage.

Learn Navigation, which you may do upon the passage, & be provided with the necessary books.

Keep a regular Diary of all transactions & events. If you describe birds, give the weights of them.

Associate very much with Indians of all Nations, for you will learn something from each.

In all your conversations with Indians be sure never to tell them that which is not true, nor even such truths as they will not readily believe and you cannot prove. In your dealings with them, never cheat them by imposing bad articles upon them. Should any article break through the fault of the maker (a thing that often happens amongst low priced hardware) exchange it for a new one.

Learn to do every thing with your own hands, that you may know when others do their work well. It is a false idea that a Gentleman disgraces himself by making use of his Limbs, for those things only are disgraceful which are contrary to the Laws of God & Man.

Spread & scrape all the Furs, which you kill, yourself, and write your name close above the tail, that the Company may distinguish yours from others; for you should take all opportunities of showing your excellencies without appearing to do it designedly.

PART ONE

"Observe Everything and Carefully Enter All": An Introduction to Labrador and the Life of Captain George Cartwright

Originally published on a subscription basis at Newark, England, the *Journal* is a unique and important source of information on eighteenth-century life on the Canadian coastal frontier. It stands alongside early Moravian accounts and British government papers in providing an unusually detailed body of material on the early colonial period and on Labrador history in particular. Its level of detail and regularity of observances in such matters as weather conditions and environment align it with the eighteenth-century art of maintaining ship's logs (Story 1980). It is therefore unfortunate that the *Journal* is not more readily available. Copies of the original folio edition are rare and costly. Facsimile copies of the complete journal are available through the Canadian Institute for Historical Microreproductions (LAC). Online it can be viewed at www.digitalbookindex.org. An abridged version was published in 1911 by C.W. Townsend as *Captain Cartwright and His Labrador Journal* (Boston). This too is a rare book but has recently become available in a reprint edition (Townsend 2003). Cartwright has also been granted a well-deserved reincarnation by John Steffler in a work of fiction entitled *The Afterlife of George Cartwright* (1992).

The new Cartwright papers supplement the *Journal* in forming an important collection of early ethnohistorical and technical information. Described in greater detail in the section entitled "A Description of the New Cartwright Papers," all are in Cartwright's handwriting and fall into four categories:

1)   A lengthy manuscript entitled *Additions to the Labrador Companion* (hereafter referred to as *Additions*), best described as an instruction manual or technical primer for prospective Labrador entrepreneurs. It includes entries on early sealing techniques, salmon fishing, hunting, trapping, and construction of buildings for Labrador;

2)   Two further manuscripts, one a long, alphabetically ordered list of materials to bring to Labrador entitled "Index for Goods," written into a narrow notebook, the other the already quoted "Memorandum for W";

3)   A collection of correspondence written by and to Cartwright, as well as nineteenth-century Cartwright family letters that mention George Cartwright;

Fig. 3   Devil's Table, one of two basaltic outliers at Henley Island, Chateau Bay.
LAC, C-065059; photo by W. Notman 1909

4)   Several loose papers that relate to the topics covered in the documents listed above but written or drawn at different times. They
include building plans for houses, a list of Cartwright's costs of
publishing the *Journal*, a map of Makkovik Bay, and equipment
and price lists.

In subject matter the papers hold closely to the topics that Cartwright
was inclined to write about in the *Journal* – hunting, trapping, sealing,
fishing, and the conscientious betterment of life in a difficult climate.
While the *Journal* introduces us to the yearly round of activities and concerns of an eighteenth-century entrepreneur, the new papers expand
on the technical aspects of these endeavours. We are given, for example,
information on how seals were trapped, on how nets were placed, which
nets served best, the cost of materials, and even the number of personnel.
Cartwright has left us descriptions of how to set different types of traps for
catching fur-bearers and the best way to build a house in Labrador, among
other matters.

With a few exceptions (i.e., Borlase 1993, 1994), much of the historical information for Labrador is still in the form of archival documents,

hard-to-find rare books, or academic research. The following chapter presents a history of southern Labrador that, while based on sometimes obscure source material, outlines precontact and contact period Aboriginal occupations, early European influences, and, in increasing detail, the later French regime and subsequent English expansion in south-central Labrador (figure 3).

CHAPTER TWO

Human Settlement in Labrador

George Cartwright's *terra familiaris* was the three-hundred kilometre stretch of coastline between Chateau Bay and Sandwich Bay. Despite its long history of human habitation, this coast, forming part of Canada's tenth province, remains as little known as Cartwright himself and has appropriately been described as "the unknown Labrador" (Kennedy 1995). The coastal landscape as seen from both the air and water is distinctive and often forbidding. The mouths of forested inlets that cut deep into the interior are linked to one another by a steep and rocky outer shore that prompted Cartwright to observe that "of all the dreary sights which I have yet beheld, none ever came up to the appearance of this coast" (C.11.7.75).[2] The Labrador Sea is buffeted by onshore and offshore winds against which the many islands offer little protection. Within the deep inlets, however, summer days can be unexpectedly hot and languid but pitched to the keening of blackflies. Over the millennia, humans have inhabited both the interior of the Labrador peninsula, the inner coast, and the outer coastal islands.

THE FIRST INHABITANTS

Labrador has been inhabited for thousands of years. The first people to live in the Strait of Belle Isle and along the south-central coast were Amerindian groups who migrated northwards from the Gulf of St Lawrence region with the retreat of the last glaciers about nine thousand years ago. During a three-thousand-year period these Archaic Indian peoples, as they are called by archaeologists, not only spread northwards along the coast to northernmost Labrador but their sites have also been found in the interior of Labrador and on the island of Newfoundland. At about 3000 Before

Present (BP), a new and different people entered northern Labrador from the eastern Arctic. These were the earliest Palaeoeskimo, an ancient Arctic culture. Two significant and separate waves of Palaeoeskimo peoples came thereafter, the Groswater and the later Dorset Palaeoeskimo. Each spread southward quite rapidly, in archaeological time, and populated all of the Labrador coast as well as much of coastal Newfoundland until approximately eight hundred years ago. During this period of Palaeoeskimo presence, Amerindian groups continued to inhabit the interior and inner coastal regions but were less conspicuous on the outer coast.

The period between 2,500 and 1,800 years ago was a time of flux among Amerindian cultures throughout the Gulf of St Lawrence, as evidenced by changing tool styles, new and diverse raw materials, and sites that suggest considerable human movement. The Amerindian peoples who came to inhabit Labrador following this period of change were the ancestors of today's Labrador Innu. Their way of life centred on securing caribou for skins, bones, fat, and meat, but small mammals such as beaver and porcupine were also key resources, as were trout and salmon. The precontact Innu lifestyle involved travel over great distances across the Labrador-Quebec peninsula, using canoes in summer and snowshoes and sleds in winter. These groups had southern Labrador to themselves from approximately 1250 BP to the 1500s. Remains of their small campsites are found in many bays and inlets not far from the mouths of salmon rivers and along river systems throughout the interior. Sustained Innu contact with Europeans began with the appearance of European whalers in the St Lawrence region in the mid-sixteenth century, and by the late seventeenth century Innu had developed relatively close ties with trade and mission stations along the St Lawrence.

The Thule Eskimo, a third important wave of Arctic-adapted peoples unrelated to the Palaeoeskimo, arrived in northern Labrador from the eastern Arctic around six hundred years ago. They were highly mobile, travelling by dogsled in winter and kayaks and the larger boats called umiaks in summer, and possessed all the other hallmarks typically associated with Arctic peoples, such as harpoons, soapstone lamps, and the half-moon shaped women's cutting knives known as ulus. There is archaeological evidence to suggest that their appearance may have displaced Amerindian groups from the coast and into the interior. By the late 1600s, the descendants of these early Thule, who are also the ancestors of today's Labrador Inuit, had spread along nearly the entire Labrador coast. They were involved in regular trade with early European vessels and, like the Innu, experienced some of the longest continuous contact with Europeans of any Aboriginal group in North America.

EARLY INNU AND INUIT CONTACT WITH EUROPEANS

Contact between Europeans and Innu groups was underway in the Gulf of St Lawrence by the mid-1500s, and growing evidence indicates that Inuit were trading in the Strait of Belle Isle by the late sixteenth century. The number of European vessels fishing in gulf waters rose exponentially and continued to rise for centuries. In 1510, fifty vessels from England, Portugal, and France were fishing on the Grand Banks, and by 1527 that same number was represented by France alone (Biggar [1901] 1972, 19). Three hundred years later the vessels fishing in these waters annually would number in the thousands.

In 1534, Jacques Cartier landed at Blanc Sablon, where he noted not only that "much fishing is carried on there" (Cook 1995, 8) but that Amerindian groups were accustomed to trading with Europeans, with both sides following established trade practices such as setting fires close to shore to signal an ad hoc trade site (Biggar [1901] 1972; Trudel 1978; Cook 1995). Basque whalers at stations between the Tadoussac region and the Strait of Belle Isle carried on regular trade (Barkham 1976, 1978, 1980, 1984; Vera et al. 1986). For example, in 1542 the crew of a Basque vessel from Orio exchanged axes, knives, and other items for caribou and seal skins in the Strait of Belle Isle (Barkham 1980), while the expedition of Micheau de Hoyarsabal overwintered somewhere along the St Lawrence in 1586–87 in order to trade (Turgeon 1994). However, it has also been documented that Inuit pilfered early Basque whaling stations as well as later French merchant posts, leading to somewhat hostile relations by the 1700s (Barkham 1980; Clermont 1980; Martijn 1980; Rapport 1922–23; Stopp 2002).

France's interest in the resources of Labrador had begun in earnest in the late sixteenth century through a growing French presence in the Gulf of St Lawrence impelled by the high price in Europe of beaver pelts and other furs as well as plentiful cod and seal stocks. It was probably the practice of the French dry fishery that created the links between fishing and the earliest fur trade. Unlike the green fishery, where fish processing was conducted on board ship, the dry fishery involved splitting, drying, and piling fish at shore facilities where direct contacts with Aboriginal groups could take place (Bigger 1901). In its early years the fur trade was incidental to the fishing industry, with fishermen and captains carrying out personal trade to supplement their wages, but by the 1600s trading took on a more commercial aspect (Innis 1930).

In the seventeenth century, European presence was focused on the cod and seal fisheries along the coastal rim. There was some interest in locating a passage through the continent but mainly by the British who wished to reach their Hudson Bay posts without having to sail northwards through Hudson Strait. The first posts of Tadoussac, Papinachois, and Moisie, all opened in the 1600s, combined religion and trade. The basis of the early ties that France made with the indigenous peoples of the peninsula, they represent the integral niche that trade held in France's New World economy. In the gulf region these trade contacts with people familiar with the interior grew increasingly valuable as the territories around Tadoussac and other stations became depleted of furs.

French initiatives northwards along the coast began with the explorations of Louis Jolliet, the first to map much of the North Shore and beyond. In 1694, under order of the Comte de Frontenac, Jolliet surveyed the coast between Mingan and Zoar, situated just north of Hamilton Inlet. His accounts revealed a challenging coastline but one that was inhabited by experienced and interested Inuit traders. Jolliet's exploration effectively opened the coast for French northward expansion, which began gradually in the early 1700s (Rapport 1943–44; Delanglez 1948).

For both Labrador's Innu of Amerindian ancestry and the Inuit of Thule Eskimo ancestry, the turn of the seventeenth century marked the end of many aspects of their old ways of life because of the significant control that Europeans began to exert over traditional coastal settlement areas, increasing dependence on European material goods, and altered social structures that came about through trade, European diseases, and/or conflict.

Inuit were gradually supplanted from the Strait of Belle Isle starting in 1700, although families were still observed at campsites along the southern coast in 1743 by Louis Fornel (Roy 1942). By the 1770s Inuit had shifted entirely out of the area, primarily due to an edict issued by Governor Hugh Palliser that forbade their presence in Chateau Bay and the Strait of Belle

Isle; it was enforced by the small resident militia at Chateau Bay, by merchants, and with the help of the Moravian mission. The establishment of Moravian missions, the earliest Protestant mission in Labrador (Rollman 2002) starting in 1770 at Nain, also pulled Inuit northwards. Much like the Catholic missions in the gulf, these Moravian missions served as centres of religion and trade. They were strategic joint ventures between Britain and the Moravian Church which were intended to shift Inuit settlement away from British enterprises in the south, and away from the ongoing altercations between Europeans and Inuit in the Strait of Belle Isle (Taylor 1977; Kaplan 1980, 1983, 1985; Clermont 1980; Martijn 1980; Kennedy 1995).

Despite the government's position on Inuit settlement, many British merchants along the southern Labrador coast would have happily continued to trade with Inuit because of the lucrative nature of trading fare such as seal oil, walrus tusks, whalebone, and pelts. Moravian trade rates, however, seem to have been unbeatable, as alluded to by Cartwright in a 1773 letter to Lord Dartmouth: "The Esquimau Indians have not been near us this Summer, the great indulgences the Moravians met with from Government have enabled them to engross all the trade of these people. To our no small disappointment and loss, for we have upwards of £300 in goods on our hands which are not saleable any where else" (LAC, Dartmouth fonds, 2444).[3]

By 1700 the Innu of Labrador had already experienced a century of Catholic conversion along the gulf coast and had well-established connections with mission trade stations, moving between them and the vast interior hinterland where they overwintered, hunted caribou, fished, trapped, and raised families. With the presence of British fisheries at every major salmon river along the southern coast of Labrador, however, Innu seasonal sojourns there became less frequent.

Cartwright traded with Innu throughout his years in Labrador but never to the same extent as with the Inuit. During the nineteenth century and as late as the 1950s, when many Innu became permanently settled in communities along the gulf coast and at Sheshatshiu, they still appeared at the Eagle and White Bear Rivers in Sandwich Bay to meet or trade with settlers and to fish (PC 1927, 1238; Tanner 1947; Armitage 1991; Mailhot 2001; Armitage and Stopp 2003).

Most Innu trade contacts in the 1800s, however, were at North West River in Hamilton Inlet and especially at posts along the gulf coast. Attempts to attract Innu to interior Hudson's Bay Company posts at Winokapau and Fort Naskapi in the mid-1800s, effectively to waylay them before they reached the posts of the gulf traders, were of limited success for a number of reasons, not the least of which were that many Innu families chose to return to the gulf posts for Christian observances and to meet extended family.

EUROPEAN EXPANSION IN SOUTHERN LABRADOR

War with Britain at the end of the seventeenth century and the resulting Treaty of Utrecht in 1713 consolidated France's holdings in the Atlantic region. France was allowed to retain portions of the Island of Newfoundland, all of Labrador, and Cape Breton Island. The French occupation of much of Labrador took the form of land concessions issued by the king to those recommended by the governors or *intendants* of the colony. Initially, large tracts of land were granted in perpetuity, but by the middle of the eighteenth century, concessions became smaller, with land grants covering four to five leagues (roughly twenty to twenty-five kilometres) of shoreline and for periods rarely exceeding nine years.

The holders of these concessions were entitled to exclusive use of the fishing grounds and had the right to compete with seasonal cod fishing vessels from France. Grant holders were expected to carry out a cod and seal fishery, trade with Aboriginal groups, and establish permanent premises and occupy them year round (Roy 1940). Between the mid-seventeenth and the mid-eighteenth century, a string of properties was granted between Mingan and North West River.

Following Jolliet's explorations, the first land grant beyond the Strait of Belle Isle was the vast concession given in 1702 to Augustin Le Gardeur de Tilly, Sieur de Courtemanche, an officer and active merchant. It began at the Kegaska River on the Quebec North Shore, had its base at Fort Pontchartrain in Brador, and extended as far as Hamilton Inlet (Esquimaux Bay) (Trudel 1978, 105; Niellon 1996, table 6.4).[4]

Few merchants actually carried out business beyond the Strait of Belle Isle, and Courtemanche's activities never extended far beyond the Bay of

Brador. By the 1740s, however, the coast had been sufficiently carved up to allow more operations, and the earliest functioning posts beyond the Strait were at Chateau Bay, Cape Charles, and Hamilton Inlet. The Chateau Bay concession was first granted to Louis Bazil in 1735–36 and later leased to Louis Fornel and partners. Fornel was still there in 1743, sealing and trading with Inuit, but the post passed to other grantees shortly thereafter. The Cape Charles concession was first granted in 1735 to Antoine Marsal, who built a sealing post there "between the islands and the mainland." In 1743 it was still the most northerly European post in Labrador. Fornel applied for the Esquimaux Bay concession in July 1743, and it was still being operated by his widow in 1762 (White 1926, 9; Roy 1940, 91; Roy 1942).[5]

The new British territories that came with the Treaty of Paris were the outcome of several years of petitioning by British merchants such as the Bristol group known as the Society of Merchant Venturers, who wished to have rightful access to northern Newfoundland and Labrador. After the treaty came into full effect in the spring of 1763, British and Jerseyian merchants quickly became established in the Strait of Belle Isle, founding long-lived and profitable cod-fishing and sealing posts at the sites of former French stations (Crowhurst 1969; Thornton 1977, 1990; Ommer 1991; Niellon 1996).

British rule, however, introduced an altogether more restrictive approach to economic enterprise than had been in place during the previous French regime. The so-called free fishery imposed by Governor Hugh Palliser's 1765 edict, "Regulations for Establishing a British Fishery for Cod, Whales, Seals, and Salmon on the Coast of Labrador" (CO 194/18, 181–2), banned residents of Newfoundland and New England from fishing in Labrador waters, and prevented land ownership and year-round rights to all British merchants, regardless of whether they had capital investments such as wharves or buildings. Palliser's intent was to allow merchant houses from Britain's West Country, whose crews sailed to Newfoundland and Labrador waters each spring, competitive access to the best harbours without having to compete with those who could more easily reach these harbours or remain over the winter, which included Newfoundlanders, New Englanders, and Canadians, or Canadiens, who were the experienced, year-round residents of the Quebec North Shore. New England vessels in particular caused many problems for Palliser. They undermined British commerce with the offer of cheaper goods, took over shoreline that was by law designated for merchant use, persecuted the Inuit, and also interrupted the French fisheries in the Gulf of St Lawrence and on Newfoundland's French Shore.

Under the free fishery regulations, the first ship to arrive in a Labrador harbour at the start of each year's fishery had full rights to that place for the remainder of the season as well as to any existing premises or equipment. However, the following year each vessel once again had to compete for prime locations. Needless to say, competition for fishing berths reached vehement proportions under this law.

The conditions of British protectionism involved a further set of hurdles whereby any goods bought, traded, or sold in Labrador had to be cleared through British customs or be of British manufacture. Cheaper merchandise from nearby markets in Quebec and New England was not allowed. Crewmen had to be of British origin, which effectively eliminated the most experienced crews, namely the Canadiens who were accustomed to Labrador conditions and knew how to seal and to fish for salmon and cod. Hiring a crew in Britain introduced a range of costs that included transport and time lost to training, acclimatization, and inexperience.

The merchants who suffered the greatest losses at the time of British takeover were those who had established themselves beyond the Strait of Belle Isle just before the Treaty of Paris took effect. Such was the case of Daniel Bayne and William Brymner, Quebec-based merchants with a four-year grant from Governor Murray dated 26 April 1763 for Cape Charles, including the salmon fishery on the St Charles River formerly held by Antoine Marsal. Bayne and Brymner were still in business in 1765 when the newly appointed Governor Palliser shut down their operations because of their non-compliance with British fishery regulations. They possessed goods purchased from Quebec and not Britain, operated a year-round premise for cod fishing and sealing at Cape Charles, and had hired a Canadien crew member (CO 194/18, 29–32; Rothney 1934). In 1767 a merchant by the name of Captain Nicholas Darby with over twenty years' experience in the Newfoundland trade and premises at Forteau and Cape Charles also had his equipment and oil confiscated on the grounds that he had hired three Canadiens to assist his inexperienced British crews (CO 194/18, 82–4). Not unexpectedly, Labrador merchants tabled several petitions before the British government in protest of the free fishery.[6] The right to exclusive property was finally recognized in a proclamation issued by Governor Shuldham in 1773 (CO 194/31, 15, 36) and was also recognized in 1774 through the Quebec Act, which brought Labrador under Quebec jurisdiction.

With the construction of a small military fortification called York Fort at Chateau Bay in 1766, and the apparent protection it afforded from errant New England whaling vessels, privateers, and Inuit, British entrepreneurs gradually expanded northeastward (CO 194/27, 262; PC 1927, 1005; Lysaght 1971).

The final paragraphs of this section focus on the region eventually inhabited by George Cartwright, although neighbouring areas such as both shores of the Strait of Belle Isle possess rich and detailed early histories. The first British operations beyond the Strait of Belle Isle were at Chateau Bay and at Cape Charles, taken over from Bayne and Brymner in 1765–66 by Nicholas Darby. Over a four-year period, Darby experienced a number of discouragements that included the confiscation of goods mentioned above but also the murder of one of his men in 1767 by three other crew members (PC 1927, 1364), the destruction of equipment by Inuit, and the refusal by crews to work during the winter seal fishery (CO 194/18, 82–5; also Moravian Mission Paper 1962, 37). The further killing of three of his crew by Inuit in 1767 resulted in soldiers being dispatched from York Fort under the command of midshipman Francis Lucas. In the ensuing fight between soldiers and Inuit, twenty Inuit men were killed and three women and six children were captured and taken to the fort. These included the Inuit woman Mikak and her son Tutauc, who were taken to England along with another Inuit boy, Karpik, by Lucas; Mikak eventually came to play a role in the acceptance of the Moravian church among northern Inuit populations (CO 194/28, 25; PC 1927, 1364; Anick 1976, 630; Taylor 1983, 1984).

This incident forced Darby to abandon the Cape Charles posts, which lay unused for two years. In 1769 the Bristol merchant firm of Jeremiah Coghlan and Thomas Perkins reinstated a salmon fishery at St Charles River, and a year later George Cartwright arrived there in partnership with Perkins and Coghlan, along with the above-mentioned Francis Lucas, who had retired from the navy as a lieutenant. Cartwright's life is described below, but it can be noted here that he took over Darby's Cape Charles sealing post and the St Charles River salmon fishery. In a series of moves between 1770 and 1786 he became the first merchant to expand into the harbours beyond Cape Charles, building his main stations at Lodge Bay, Stage Cove, Sandwich Bay, Paradise, and Isthmus Bay. He also established smaller outposts for sealing, cod, salmon, and winter trapping runs along the coast (Cartwright 1792; *DCB* for "Jeremiah Coghlan," "Francis Lucas").

Although he spent intervals in England between 1770 and 1786, Cartwright's main residences in Labrador were Ranger Lodge at Lodge Bay (1770–74), Stage Cove (1774–75), Caribou Castle at today's community of Cartwright (1775–78), Isthmus Bay (winter of 1778–79), Paradise River (season of 1783), and again Isthmus Bay (1785–86). Other British merchants were quick to follow his lead. As table 1 shows, by the 1780s British expansion along the south-central coast was all encompassing. Most merchants had cod-fishing vessels on the Labrador Sea, all the major rivers had salmon posts at their mouths, and all prime sealing locations between Chateau Bay and Hamilton Inlet had been claimed.

The onset of the American Revolutionary War (1775–83) brought some setbacks to enterprise along this coast. American privateers attacked communities and merchant stations throughout the eastern seaboard and the Gulf of St Lawrence. In 1778 alone the fishery along the coast of Labrador was brought to its knees by successive privateer attacks, forcing some merchants, Cartwright among them, into financial difficulties. The situation prompted Governor Edwards to provide arms to various harbours in the absence of a defensive force, including to Cartwright at Sandwich Bay (CO 194/33, 31, 128, 130; CO 194/34, 34, 54, 63; also Ommer 1991, 26). These pressures had little long-term effect on the growth of industry, however, and merchants continued to maintain crews along the coast.

The heyday of the cod fishery came in the early decades of the 1800s. In the Strait of Belle Isle, this fishery was largely controlled by merchants from Jersey who had strong economic ties to England and maintained posts between Brador and Red Bay. Just north of the Straits, between Chateau Bay and Sandwich Bay, large merchant concerns with ties to the West Country or Newfoundland participated in a fishery that by 1860 had taken on immense proportions, bringing fifteen hundred to eighteen hundred schooners to the shores of Labrador from Newfoundland alone, bearing fifteen thousand to twenty thousand men, women, and children for the summer season. Inuit families from Spotted Islands, Hamilton Inlet, and northwards also worked in the fishery. Families lived in sod houses that are still visible as grass-covered mounds along the length of the southern

Table 1: Eighteenth-Century Merchants in South-Central Labrador

Date	Location	Purpose	Merchant Name
1735–1750	Cape Charles	cod. sealing	Antoine Marsal
1736	Chateau Bay	cod, sealing	Louis Bazil
1737–at least 1742	Chateau Bay	cod, sealing	Louis Fornel leased from Bazil
1749–1754	Chateau Bay	cod, sealing	Gaultier
1749–1753	Cape Charles	cod, sealing	Capt. de Bonne (Baune?)
1753–58	Cape Charles	cod, sealing	Antoine Marsal
1754–?	Chateau Bay	cod, sealing	Gaultier deeded rights to Sieur de Lanaudière and Ches. Gilbert
1758–1763	Cape Charles	cod, sealing	At Marsal's death, concession granted to executors for remainder of 9 year term (term began in 1753); no evidence of use
1763	Cape Charles	cod, sealing	Daniel Bayne & Wm. Brymner
1771–1774	Temple Bay & various parts of Chateau Bay, Bad Bay (and its Seal Island), St Peter's Bay, Seal Island at Cape Charles, Mary's Hbr., Hoop-Pole Cove, Fox Harbour, Niger River	sealing, cod, salmon	John Noble & Andrew Pinson
1765	Chateau Bay	sealing	Jeremiah Coghlan & Thomas Perkins
1765–1767	Cape Charles	cod, sealing	Nicholas Darby
1767–1770	Charles R.	salmon	Coghlan, & Perkins, & George Cartwright & Francis Lucas

Date	Location	Purpose	Merchant Name
1770	Cape Charles	sealing	Cartwright
1770	Charles R.	salmon	Cartwright
1772	Cape Charles	sealing	Noble & Pinson
1772	Charles R.	salmon	Noble & Pinson
1772	Colleroon (St Lewis R.)	salmon	Cartwright
1773	Colleroon (St Lewis R.)	salmon	Noble & Pinson
1773	Rivers between Alexis R. and Porcupine Bay	salmon	Coghlan
1774	Henley Hbr., Seal Island & Bad Bay	sealing, cod	Slade
1774	Henley Hbr., Alexis R., Spear Hbr.	sealing, cod, salmon	Coghlan
1774	Indian Isl., Little Tickle (Cape Charles), Gilbert R.,	cod, sealing, salmon	Hooper
1774	Stage Cove, Charles R., Port Marnham, White Bear Sound, Enterprise Tickle, Square Islands	sealing, cod, salmon	Cartwright
1775	Charles R.	salmon	Cartwright
1775	Fox Hbr. & Lance Cove	sealing	Noble & Pinson
1775	Alexis R.	salmon	Coghlan
1775–1784	Paradise R.	salmon, trapping	Cartwright
1775	Eagle R.	salmon	Cartwright
1775/76	White Bear R.	salmon	Cartwright
1775	Alexis Bay area	salmon	Hooper (planter named Baskomb)
1775	Mary's Hbr.	salmon?	Noble & Pinson

/continued on next page

Date	Location	Purpose	Merchant Name
1775	Cape Charles & Great Caribou	unknown	Hooper
1775–1778	Cartwright	main station	Cartwright
1776–1784	Eagle R.	salmon	Cartwright
1776	Partridge Bay	furring, salmon	Coghlan
1776	Sandy Hill Cove, Porcupine Bay	furring, salmon	Coghlan
1777	Coast between Alexis R. and Porcupine Bay	furring, cod, salmon	Coghlan
1777	Table Bay	sealing, fur	Coghlan (planter named Wrixon)
1777	Groswater Bay	furring, cod, salmon	Coghlan (crew of Charles Hellins)
1778	Grady Islands & Curlew Hbr., Great Island	sealing	John Gready (planter for) Cartwright
1778/80	Table Bay	sealing, furring	Cartwright
1778	Table Bay	sealing	John Wrixon (probably working for J. Coghlan)
1778	Sand Hill Cove	salmon	Cartwright
1778/79	Porcupine Bay	furring	Coghlan
1778–1780	Porcupine Bay	cod fishery	Phippard
1778–1790	Sand Hill Cove	furring	Coghlan
1783	Sand Hill Cove	sealing	Cartwright
1783/84	Sand Hill Cove	sealing	Noble & Pinson
1784/85	Sandwich Bay rivers	salmon	Noble & Pinson
1785/86	Muddy Bay (Sandwich Bay)	sealing?	Nicholas Gready
1785	Groswater Bay	trade, sealing	M. Marcoux
1785/86	Groswater Bay	?	"some Englishmen"
1785/86	Groswater Bay	furring	M. Demoitie et al.
1786	Hawke Isl. (Eagle Cove)	sealing	Mr. Hyde
1786	Hawke Bay area	cod	Arthur Thomas Slade

Date	Location	Purpose	Merchant Name
1786	Square Isl.	cod	Netlam Tory
1786	Seal Islands (Isl. of Ponds)	sealing	Marcoux
1785	Venison Hbr.	sealing	Mr. Hyde
1786	Eagle Cove (Hawke Isl.)	cod	Mr. Hyde
1786	Square Island	cod	Tory

Note: Dates given are either the earliest known date of presence or known years of use. Sources: Cartwright 1792; White 1926; PC 1927; Rothney 1934; Roy 1940; Anick 1976; Kennedy 1995; also Dartmouth fonds, CO 194

Labrador coast, and many of these seasonal visitors remained in Labrador (Gosling 1910, 315; Whiteley 1969, 1977; Thornton 1977; Kennedy 1985; Charest 1998; Stopp 2002).

THE EARLY PERMANENT SETTLERS OF SOUTHERN LABRADOR

Cartwright's time in Labrador also marks that of earliest permanent European settlement between Cape Charles and Sandwich Bay. By the late 1700s small numbers of British men had made Labrador their home and had begun families with Inuit women, blending Inuit food, clothing, technology, and environmental knowledge with European religion, language, and economy (Kennedy 1995). One of the earliest references to a European resident family is found in the *Journal* and concerns Cartwright's foreman and boat-builder, John Baskem, who arrived in 1771 from Fogo with his wife (unnamed) and two children. Baskem built himself a house in Port Marnham, on the north shore of St Lewis Inlet, and worked for Cartwright for many years. In 1779 he deserted from Cartwright's employ to become an independent planter, taking some of Cartwright's possessions. Intent upon retrieving what was his, Cartwright came to St Lewis Inlet only to find himself chastened by the sight of Baskem's wife and children living in complete poverty, even without shoes. Through written deed he gave the family his buildings and equipment in the Port Marnham area as well as salmon racks and cribs at Cape Charles (C.24.10.71; 13.9.79).

Kennedy's (1995, 1996) studies of early settlement of the coast noted a way of life characterized by such poverty and debt to merchants, but with a seasonal rhythm that remained a persistent lifestyle from the late 1700s until the cod moratorium of 1992. Families carried out a series of shifts, also known as seasonal transhumance, between the outer and inner coast that brought them closer to each season's resources. In many respects this way of life finds its roots in early merchant resource harvesting, whereby crews made these same moves between the coast and the inner bays and forests. During the warm months, families lived in coastal communities to fish a variety of saltwater species such as cod, caplin, and herring and to hunt seals during early spring and early winter. In late October there was a shift to the heads of the inlets for the cold season where the last of the berries could be picked, a wood supply maintained, and hunting and trapping carried on. These winter settlements were also often close to early summer salmon runs.

There have been few chroniclers of early winter life other than Cartwright, and it has been aptly observed that "once families left the outer coast each fall for their winter quarters, it was as if they became invisible" (Kennedy 1996, 25). A brief mid-nineteenth century record of this lifestyle was made in 1859 by a visitor to Chateau Bay who had spoken with a resident there named Mr Clark: "According to the custom of those who live permanently in Labrador, Clark and a few of his neighbors remove, in autumn, to the evergreen woods along the streams at the head of the bay, and spend more than half of the year in hunting and sealing, and getting timber and firewood for the summer" (Noble 1861, 268; also de Boileau 1861; Butler 1878). An 1840s visitor also noted "they invariably carry all their property with them, in their annual migrations to and from their summer residence" (Davies 1843, 88). Many of today's older residents in south-central Labrador have vivid memories of annually shifting their belongings, their eiderdown mattresses, dishes, woodstoves, and even windows, back and forth between homes at the coast and those up the bays.

In 1783, Cartwright recorded two of the southernmost Inuit communities of that time, at Spotted Islands and Huntingdon Island (c.5–8.9.83). Inuit presence along the south-central coast did not disappear altogether after the influx of early English merchants, even making a resurgence in the early nineteenth century with the growth of mixed households. Bishop

Feild, who visited Sandwich Bay in 1848, found a largely Anglican population there, "many pure Esquimaux, but the majority Anglo-Esquimaux," and remarked that they were well versed in the Anglican catechism, despite never having had a visit from a representative of that church, and many could read and write English (Feild 1849, 19). Early families such as these constitute the ancestors of today's population along the south-central coast (Tanner 1947; Borlase 1993, 1994; Fitzhugh 1999; Buckle 2003).

Southern Labrador has been settled by different peoples over time, and the reasons for culture change may be found in factors such as human migrations, human ingenuity, the impacts of climate on settlement and food resources, the impacts of European diseases, politics, and economic needs, and recently the impacts of overfishing in the North Atlantic. Perhaps most persistent throughout the region's long history has been the annual passing of the harp seal herds northwards and southwards along the coast, and the annual arrival and spawning of salmon, which have provided meat, skins, and bones, and a measure of certainty for all peoples who have lived along the coast.

An Account of George Cartwright's Life

Cartwright tells us in the introduction to his *Journal* that he was born on 12 February 1739, which puts his birthdate on 23 February 1739 in the Gregorian calendar adopted by Britain in 1752. He was born into an old family at Marnham in the county of Nottinghamshire, England. His formal education ended at the age of fifteen, when he became a cadet at the Royal Military Academy at Woolwich, eventually serving in India until around 1758, then Ireland, where he was promoted to lieutenant in 1759, and as a captain in Germany until 1762. Between 1762 and 1766 he retired from the army on half pay.

His introduction to Newfoundland and Labrador came in 1766 when he sailed there with his brother, John Cartwright, first lieutenant of the *Guernsey* under the colony's governor, Commodore Hugh Palliser. On their return journey in the autumn of 1766, the brothers made up part of an historic meeting in St John's harbour where the *Niger*, the *Grenville*, and the *Guernsey* were anchored for a time. These ships brought together Governor Palliser, Joseph Banks, James Cook, Thomas Adams, who was charged with building York Fort at Chateau Bay, and Michael Lane, who later completed the cartographic work begun by Cook in Newfoundland.

Upon his return to England, Cartwright briefly held a captaincy in the 37th Foot in Minorca but within the year came home to recover from malaria. In 1768 he made a second voyage to Newfoundland with his brother John, whom Palliser had charged with an expedition into interior Newfoundland to establish friendly relations with the elusive Beothuk Indians. Palliser's and John Cartwright's strong humanitarian positions *vis à vis* indigenous peoples were the inspiration for George's own beliefs, which influenced him in his relations with Inuit and Innu throughout his years in Labrador and also led to a 1784 deposition to the Colonial Office in support of a Beothuk land use area between Gander Bay Point (probably today's Dog Bay Point at the mouth of Gander Bay) and Cape St John (CO 194/35, 336–7, *DCB* "Cartwright"; Whiteley 1969; Lysaght 1971; Marshall 1996, 113–15).

In 1770 Cartwright returned to Labrador to establish his first merchant venture at Cape Charles, in partnership, as noted, with three men with previous Labrador experience: the merchant partners Jeremiah Coghlan and Thomas Perkins, and Francis Lucas, who had served at York Fort in Chateau Bay.

Despite the great energy behind Cartwright's efforts between 1770 and 1786, he nevertheless suffered nearly continuous business setbacks, generally due to forces beyond his control. These began in his first year, with the death of Lucas in a shipwreck, followed by a fire in 1772 that destroyed the buildings at his first rooms at Lodge Bay. He had frequent changes in partnership, probably instigated by the largely unprofitable nature of his business ventures. He experienced first hand the ill effects of the free fishery when his St Charles River salmon post and Cape Charles stage were taken over in 1772 by his greatest competitor, the merchant establishment of Noble and Pinson, who also took over his salmon post at the head of St Lewis Inlet in 1773. He successfully petitioned Governor Shuldham for the return of the two salmon posts, reclaiming them in 1774 (LAC, Dartmouth fonds, 2442).

Other calamities for Cartwright included a 1778 privateer attack at Sandwich Bay that resulted in losses of more than £14,000 sterling. In this case, his one consolation may have been that all merchants along the southern coast were similarly affected that summer. In 1781 his supply vessel *The Countess of Effingham* was dashed to pieces on the shores of Trinity Bay. A new schooner built in 1781 at Paradise, Sandwich Bay, arrived in England no longer seaworthy and with much of its cargo thrown overboard. Following repairs, it was captured in 1782 by an unnamed enemy, probably privateers. A third supply vessel, *John*, sank in 1783 with a cargo of furs and whalebone.

Cartwright placed great hopes of turning his poor fortune on the discovery of a vein of ore in Sandwich Bay. But when it proved to be of no value, "all my expectations from it were blasted, and I clearly foresaw inevitable ruin, waiting my return to England" (C.12.9.83). The following year he declared bankruptcy, and his assignees in London, Robert Hunter and William Sharp – who were, he claimed, also working for Noble and Pinson – sold the largest and last of his assets, the rooms at Sandwich Bay, to that company for a pittance of £250 (LAC-RG1, letter from Cartwright to E. Nepean, 4 January 1787).

Cartwright's *Journal* is rich in many respects, presenting an account of the daily affairs of this retired military man who in midlife set out to become a merchant in a relatively forbidding corner of the British Empire. It contains ample evidence of a man with remarkable energy and focus, which was noted by contemporary observers such as Robert Southey who described him as having "strength and perseverance charactered in every muscle" (Lysaght 1971). The *Journal* also contains accounts of the activities of his crew, his relations with Innu and Inuit, some of their customs, and the effects of rival merchants upon his business.

Perhaps most persistently, he recorded his nearly daily hunting exploits. Prowse (1895, 599) wryly observed that "every animal killed is noted in this immense diary; even the most inveterate sportsman would tire a little after perusing sixteen years of this daily work of slaughter." Sportsman Cartwright also was, with a changing array of bloodhounds, foxhounds, greyhounds, pointers, spaniels, and Newfoundland dogs, along with several types of firearms. But it was sport with a purpose, as fresh meat was a preventative against disease and a significantly cheaper source of food for his crews than pickled foodstuffs from Britain. Perhaps the strongest motivation for continuous hunting was that he had "so great a dislike to salted meat, that I would rather eat any animal whatsoever that is fresh, than the best beef or pork that is salted" (C.22.2.76).

Out of sheer necessity he became an adept household medic, administering arcane remedies for a variety of illnesses, pulling teeth, amputating, blood-letting, and mending broken bones. He became a midwife during his first winter in Labrador, successfully delivering his servant Nanny's "stout boy" but declaring, "I was obliged next to act as nurse, and take the child to bed with me; neither of which offices do I wish ever to resume" (C.12.2.71). Over the years he delivered several more children, including the breech-birth twins of Nooquaschock, an Inuit woman who worked in his household. David Scully, the twins' father and a crewman of Cartwright's, eventually took his wife and children back to Ireland (C.18.10.76).

Cartwright himself suffered from a variety of illnesses, especially in times of stress, and he was not averse to dosing himself from his large supply of remedies, examples of which are listed in "Index for Goods." One of his self-treatments for a painful bout of sciatica and lumbago involved taking laudanum and a laxative known as jalap, letting eight ounces of blood, taking calomel, camphor, and a substance he called guaiacum, and applying a plaster to his thighs sprinkled with cantharides (Spanish fly), which served as a blistering agent (C.22–24.6.79).

Considered in the context of his time, he was a model product of the British Enlightenment. His broad interests reflect this age of discovery and its fundamental social, economic, and intellectual shifts. Fittingly, he was an inveterate innovator, experimenting with hunting techniques, gardening, animal domestication (he brought European livestock to Labrador and raised a pet caribou), and the application of Aboriginal knowledge. He dutifully logged meteorological data that can be found in the endpapers of the *Journal*, and recorded aspects of natural history that were of particular interest at that time, such as the habits of beavers and the phenomenon of colour change in northern animals. He provided the first accounts of seasonal moulting and colour change of ptarmigan and hares, information of great interest to natural scientists such as Joseph Banks and Daniel Solander. Although he was not a scientist himself, Cartwright's observations on Labrador's flora and fauna, climate, and Aboriginal peoples were in keeping with the expanding fields of biology, anthropology, and classification of the natural world. Throughout his years in Labrador, he relayed observations and collections of plants and animals to Banks. Lord Dartmouth, governor of the Board of Trade and Plantations, also received a number of specimens from Cartwright, including a large beaver skin and the head of a caribou stag. The surgeon John Hunter received a beaver skin, while the Duke of Rutland received silver foxes from Cartwright through the merchant Benjamin Lester in 1783 (C.14.12.72; JBRP 1773–83; LAC-Dartmouth fonds, 2460–4; DRO-Lester and Garland Archive; *Lloyds Evening Post* January 1773; Dawson 1958, 203; Lysaght 1971, 79–89).

Cartwright did not consider himself a well-educated man, rather too humbly describing himself as "a faithful Journalist, who prefers the simplicity of plain language and downright truth" (C.viii). In a similar overly modest vein, he wrote to Banks that the *Journal* "may, perhaps, sometimes create a laugh, but instruct, I think, [it] cannot" (Lysaght 1971, 275). He attributed his literary skills and "little improvement" in such matters to his daily writing while in Labrador, when in fact his work went far beyond the *Journal*. In addition to his correspondence with family, in 1785 he published a 365-line poem about a year's activities in Labrador that served the dual purpose of celebrating this new wilderness while inviting British expansion (cf. Rompkey 1987). He wrote a number of position papers, or memorials, to government on such subjects as the problems of Britain's position on year-round settlement in Labrador (CO 194/31, 7–13; LAC, Dartmouth fonds, 2434–6). In 1773 he detailed cogent reasons for administering Labrador through Newfoundland rather than through a Canadian administration based in Quebec (LAC, Dartmouth fonds, 2386–93). Ten years later he presented a further epistle on the state of the Labrador trade under Quebec and the urgent need for self-government and civil justice, tentatively suggesting himself as a resident governor (ibid., 2469–78). He also maintained formal correspondence with Lord Dartmouth that contains invaluable observations on economic developments along the coast and details on his business (ibid., 2438–54, 2460–4, 2469–70).

Personal sentiments are often given short shrift in the *Journal*, and Cartwright reveals little of what he must have felt during sixteen relatively tumultuous years. Signifiers of deeper concerns are difficult to identify in his even-keeled prose, and are often touched upon only briefly and in the lightest manner, with unbridgeable gaps between "what was said and what hovered, just unsaid, between the lines" (Atwood and Pachter 1997, xi).

He did express great sadness upon the deaths through smallpox of the Inuit he brought to England on two occasions. The first group travelled with him to England in October 1772. They were Attuiock, whom Cartwright described as "an old priest," and Attuiock's youngest wife, Ickongoque, and Ickeuna, her daughter, under four years of age. There was also Tooklavinia, who was Attuiock's youngest brother, and Caubvick, Tooklavinia's wife. All except Caubvick died of smallpox just before leaving England in the spring of 1773. Full-figure portraits of Caubvick and Attuiock were drawn by Nathanial Dance in the winter of 1772–73 for Joseph Banks (figure 4).[7]

The following year, Cartwright brought a young Inuit boy, Noozelliack, to England in hopes of having him learn the English language and become his interpreter and go-between with Inuit in northern Labrador. As a

Fig. 4   Portraits of Attuiock and Caubvick, painted by Hünneman in 1792 from the full-figure portraits by Nathanial Dance (shown in Lysaght 1971; see note 7). Courtesy of the Ethnological Museum of Göttingen, Germany

first step Cartwright had him inoculated against smallpox by the surgeon Daniel Sutton. For many decades Britain had experimented with inoculation by placing powder from the scabs of smallpox onto scratches on the skin. Until Edward Jenner discovered the cowpox vaccine (1796) as a non-fatal way of developing immunity in humans against smallpox, however, there was no way of estimating correct amounts, and the mortality rate was high. To Cartwright's dismay Noozelliack died of smallpox three days after receiving an inoculation (C.15.12.73). The letter from Daniel Sutton to Cartwright, informing him of Noozelliack's death, forms part of the correspondence presented below.

Other life-changing events that Cartwright gave only mild mention included Noble and Pinson's takeover of the Cape Charles premises in 1772, the many failed sealing seasons, the loss of several vessels, the destruction of his home in Sandwich Bay, and the kidnapping of his Inuit crew by the privateer Grimes in 1778. These events, among others, are described but not dwelt upon in the *Journal*; their being noted at all seems to be the full representation of Cartwright's passion.

His housekeeper in Labrador between the years 1770 and 1779, Mrs Selby, is hardly referred to at all in the *Journal* except on occasions when she accompanied him hunting or fishing, and she remains an obscure figure. The birth of her daughter in 1779, with Cartwright acting as midwife, is noted briefly alongside that day's other events: "At day-break Mrs. Selby was taken in labour, and at ten o'clock I delivered her of a daughter. At seven Mr. Coghlan's men set off for Sandhill Cove. Daubeny and Collingham went round the traps ... and brought in a white-fox. Three men were throwing the snow out of the cook-room. Sharp frost, and clear all day" (C.21.4.79).

In a *Journal* entry a month later, however, after nine years of entries, Cartwright reveals his true relationship with Mrs Selby, when he declared "as formal a divorce between us as ever was pronounced in Doctors Common" (C.26.5.79) on the grounds that Joseph Daubeny was the child's father (but denied it). The child lived only a short time, and that summer Cartwright saw Mrs Selby safely back to England and provided her with an allowance for life. In a puzzling spirit of fraternity, undoubtedly borne of necessity, he continued to work with Daubeny.

Utterly intriguing is a shred of evidence that Cartwright may have had a son by an Inuit woman. This tantalizing piece of information comes from the Nain Moravian mission diary of 1782. The entry contains news passed to the missionaries by Inuit returning to Nain after trading in Chateau Bay and along the south-central coast. These Inuit were well-known figures in the history of Labrador, namely the Inuit woman Mikak, her husband, Abraham, and her former husband, Tuglavina, who had travelled with their immediate and extended families in a convoy to obtain goods in the south not available through the mission, such as boats and firearms. In their account to the missionaries of the people and events they had met with there is the additional information that "they had also been at a settlement of Mr. Cartwright; had seen his Son, who had expressed a great desire to pay us a visit" (Nain Diary 1782, 3 October).

It is possible that the Inuits' understanding of the true relationship between Cartwright and this boy was misinterpreted or mistranslated by the missionaries, for Cartwright could have been the guardian of an Inuit child rather than father. The probability that he might have had Inuit progeny, however, is high considering his years in that country, the close ties he held with Inuit, and the fact that relationships between European men and Inuit women were not unusual. Were this Cartwright's child by Mrs Selby or another European woman, we might have learned more of him in the diary. The fact that the Inuit were able to communicate with the boy, who expressed an interest in visiting an Inuit settlement, also suggests a native heritage (although Mikak could speak some English). There is little more that can be wrung from these all too brief words. The *Journal* contains no reference to a son except for occasional mention of "the boy," which could also refer to the apprentices that commonly made up a merchant's crew.[8]

In 1784 Cartwright was bankrupt and saw no further means of continuing as a merchant. He unsuccessfully petitioned government for a source of employment, suggesting for himself the governorship of Labrador (CO 194/35, 336). In his proposal for protection of the Beothuk of Newfoundland, also submitted in 1784, he recommended that a superintendent of Indian Affairs be created and again offered his services, but this innovative and discerning idea came to nothing (Marshall 1996, 114). Despite these rejections, he nevertheless returned to Labrador for two more years to engage in salmon fishing, sealing, and a small cod fishery with a business partner, Robert Collingham.

In 1785, the rival merchants John Noble and Andrew Pinson, who had been pursuing Cartwright for many years, succeeded in taking over the very lucrative Sandwich Bay salmon fishery that included the main residence of Caribou Castle in today's community of Cartwright, the wharves, boats, and outbuildings, as well as the extensive installations on the Paradise, White Bear, and Eagle Rivers.[9] In a rare show of feeling, Cartwright lamented his losses at the hands of Noble and Pinson and never returned to Labrador: "My late possessions in Sandwich Bay, together with what goods remained there, were sold last winter, by my assignees, to Noble and Pinson; for the paltry sum of two hundred and fifty pounds: whereas, the goods alone ... were valued ... at two hundred and eighty pounds; and ... the fishing-posts and the buildings thereon, were well worth a thousand pounds ... I must confess that I cannot help feeling greatly hurt, that Noble and Pinson, who have been my inveterate enemies ever since I first came to this country, should get, for less than nothing, possessions, which cost me so much labour, to find out, and money to establish" (C.4.9.85). Even the merchant James Slade commiserated with Cartwright, referring to Pinson as a "Bird of Prey" and to Noble as only interested in "men with power," while rightly observing "they have finally benefited by your discoveries, Improvements and Misfortunes" (LAC RG1, L3L, 15042).

Cartwright sailed back to England for the last time in the late summer of 1786, but not without incident. In the port of St John's, he embarked on the brig *John* (Captain John Bartlett), sharing a cabin with a rather infamous fellow passenger, the turncoat General Benedict Arnold. In late October, five hundred miles out of St John's, the ship nearly capsized in one of Cartwright's worst experiences at sea. After repairing masts and sails, all on board were also put on water rations in the event of further storms. These came with a vengeance, pushing the ship ever northwards. When the *John* finally arrived in England in November, Cartwright learned that the water shortages among the company had been caused partly by Arnold, who had set up a scheme to trade his and Cartwright's supply of wine to the sailors in return for water for himself.

Cartwright may have held business interests in Labrador as late as 1793, seven years after returning to England. On his answering questions before a 1793 parliamentary committee on the state of trade of Newfoundland, the following record was made of his remarks: "And being asked, Whether there is not a more flourishing trade carried on at Labradore than at Newfoundland he [Cartwright] said, He could only say, with respect to himself, that his trade has been very flourishing, having cleared above one hundred per cent. for the last three years" (Great Britain 1785–1808, 403). His statement undoubtedly refers to the three years preceding the 1793 inquiry, and suggests that he may have kept an investment in the Aboriginal trade through Robert Collingham. We know that Collingham was still in Labrador in 1791 because he had plans to set up a salmon fishery with the trader Marcoux at Kipokok, just south of Hopedale (*Periodical Accounts* 1797, letter from Hopedale, 7 October 1791).

Between 1786 and 1819, Cartwright remained active in Labrador matters, publishing his three-volume diary, completing a manuscript entitled *The Labrador Companion*, writing the material presented here, and unsuccessfully petitioning for a land grant for a post in Hamilton Inlet and the position of Justice of the Peace (CO 194/35; PC 1927, vol. 3, 1178–80). In the latter petition he was supported by other Labrador merchants including Netlam Tory who had taken over Cartwright's Square Islands fishing room, Henry Nicholls at Temple Bay, and Mr Ross at Brador (LAC RG1, L3L, 15042–51). He eventually accepted the position of barracks master with the Nottingham militia, which he held until 1817. He died two years later at the age of 80 and was buried in Mansfield, Nottinghamshire, on 24 May 1819 (Prior 1925, 36). For a time he had lived in a building still known as Black's Head, on Broad Marsh in Nottingham (Nottinghamshire 2006). In the folk history of that area he is remembered as having pursued the ancient art of hawking, or falconry, and was locally known as "Labrador Cartwright" and "Old Labrador."

Other descriptions of Cartwright's life can be found in Lysaght's (1971) account of Joseph Banks's summer in Labrador, in Kennedy's (1995) study of the southern Labrador coast, and in George Story's entry on Cartwright in the *Dictionary of Canadian Biography* (*DCB*). The richness of his life and the many papers that have survived allow ample room for yet another biographical account. According to his niece, Frances Dorothy Cartwright, he lived his entire life "possessed of uncommon vigour both of mind and body" and even a few months before his death "was busied in proposing to the Hudson's Bay Company, various plans and contrivances for hunting, &c. and nothing but increasing infirmity prevented his offering his services to put them in execution" (Cartwright [1826] 1969, 159). Although we cannot be certain, these "various plans and contrivances for hunting" may have included the papers presented here.

CHAPTER FOUR

A Description of the New Cartwright Papers

The new Cartwright papers are made up of a manuscript entitled *Additions to the Labrador Companion* as well as shorter texts, correspondence, and miscellaneous material on loose sheets. The papers reaffirm the image of Cartwright as an energetic chronicler and correspondent and, as with his *Journal*, find their place in period writings through their themes of innovation and efficiency, developing technologies, and classification.

As its title implies, *Additions to the Labrador Companion* was additional material to complement a previously written manuscript evidently entitled *The Labrador Companion*. We know nothing of this document except for hints in *Additions* that suggest that it was a lengthy volume and not dissimilar to *Additions* in subject matter and style. For example, on page 7 of *Additions*, Cartwright drops a clue when he instructs "Throw a Bridge across the River consisting of two sets of Beams supported with Posts and Shores, as directed in Page 326 of the Labrador Companion." From this passage we learn that *The Labrador Companion* included instructions of a technical nature similar to *Additions*, and that the manuscript ran to at least 326 pages. Cartwright refers to the existence of *The Labrador Companion* eight times in *Additions*, as listed in table 2.

Probably handwritten and loaned to interested parties, *The Labrador Companion* remains an intriguing lost manuscript, hopefully still to be found. Cartwright may not have pursued its publication after the arduous process of publishing the *Journal* on a subscription basis. One of the letters transcribed below, written by the Hudson's Bay Company to Cartwright in 1818, a year before his death, expressed the HBC's interest in purchasing his "Book of remarks on Labrador," which perhaps refers to *The Labrador Companion*. That he did prepare "various plans and contrivances for

Table 2: Links between The Labrador Companion *and* Additions

Page reference in Additions	Page reference in The Labrador Companion	Shared subject matter
7, 67	326	Bridge building
37	50	Building a trap house
38	17	Improvements on the marten trap
47	178	Roofing a house
93	39	Directions for trapping otter
96	242	Instructions on building houses for Inuit
97	291	Goods for Inuit trade
100	202	How to find wolf cubs

hunting" for the H BC during his final year is corroborated by his niece in the passage quoted above.

ADDITIONS TO THE LABRADOR COMPANION: A DESCRIPTION

Despite its table of contents, often the sign of a completed text, the general structure of *Additions* is clearly that of a working manuscript; entries are arranged as they came to mind rather than by topic, and some are incomplete. That it was not ready for publication is also evident in the paper that it was written on, which has the text for another manuscript on the verso face.

Additions is a sheaf of 112 pages, each leaf 25.5 cm by 19.5 cm. The first page is unnumbered and puzzling, bearing only the centrally positioned name of Cartwright's brother, "John Cartwright," and in the top left corner a short notation that reads "Weight of a Green-plover's egg 14 Drahms." This sheet may be unrelated to the manuscript but it is also conceivable that it is a dedication (on used paper) to Cartwright's brother, Major John Cartwright, who had supported his endeavours in Labrador and shared his interests in Aboriginal peoples. This top sheet is followed by two pages that list the contents of the manuscript, followed by the first page of the text itself, which is numbered "1" in the top left corner. The numbered sheets continue through to page 108, followed by a final unnumbered page of text.

As already mentioned, there is little subject organization to the entries in *Additions*. The document launches without introduction into a two-sentence notation on how to flush animals from underground. It moves to instructions for building an otter trap, and then on to a flotation device, ending 109 pages later with an entry on deer pounds. Despite the shifting themes and stream-of-consciousness approach, the transcription holds to the original ordering and pagination. (To make it easier to find entries by subject, Cartwright's contents page is reorganized by subject matter in table 3.)

The second manuscript on the back of each leaf of *Additions* is unrelated to Labrador affairs and consists of fifty-four pages of information on English countryside activities such as the art of shooting, hounds, game conservation, and fishing techniques. These entries have not been included here, but they do provide further evidence that *Additions* was a draft.

When was *Additions* written? Two pieces of information in *Additions* allow its dating. On page 25 Cartwright noted that he had not seen Venison Harbour in twenty-four years, and from the *Journal* we know that his last visit there was in 1786, during his final departure from Labrador (C.5.8.86). This indicates that *Additions* was written around 1810, a time-frame confirmed by a notation on page 48 that gives the time of writing as February 1811.

THE OTHER NEW CARTWRIGHT PAPERS

The remaining Cartwright papers fall into three categories:

1) Two manuscripts of some ethnohistoric interest. Their themes are closely tied to information found in the *Journal* and *Additions* but with additional detail.

    a) The first manuscript in this category is an alphabetically ordered "Index for Goods" that was written into a narrow notebook approximately 5 cm by 17 cm. It is both a curiosity as an early list of material goods and of ethnohistoric interest, list-

Table 3: Cartwright's Table of Contents for Additions to the Labrador Companion, *Rearranged by Subject Matter*

General hunting, birding

Any Beast in an Earth to catch   1

Otters upon a Path to catch   1

Long-winged Hawks, to teach to wait on   4

Saltwater shore to Fence   4

To Mew an Eyess hawk, the best way   6

Bears, Wolves, Deer or Foxes, to catch on a hook   40

Snow Bunting   46

To provide covering for Rub[g] places   48

Birds to catch with fish hooks   53

Magpies to destroy   53

An excellent Drag for Bear, Wolf or Fox   53

To make a ladder to rob a birds nest in a tree   55

Fox or wolf cubs to catch alive   56

Magpies & Carrion Crows to catch   57

Crows, Magpies & Hawks to poison   58

Flat, to shoot Geese upon a Shoal   58

Geese to shoot upon a flat woody shore   59 [sic][1]

Curlews, Golden Plover etc. to shoot   61

An excellent bait for Foxes & Wolves   69

Wolves & their cubs to catch   100

Bullet-moulds   106

Gun-box & its contents and also Ammunition box   107

Various structures

Back tilt – portable one to construct   2

Plan of Ditto   3

Posts or studs, bottoms of them preserve   12

Store-house Stable & Hay room plan of   32

Dwelling House plan of   33

Smoke-house & Bake-house   34

Permanent Wharf to Build   34

Quantities of Plank, Glass etc. for dwelling   35

An excellent covering for Houses   47

An entrance unto a House exposed to drift   57

Studded House to build   61

House of Boards on the Ground   63

To secure Houses from Wind   66

An expeditious way for fixing Beams & Rafters, made of Planks   70

To prevent leaks in a boarded up house   70

To fix the Stancheons of outer walls or Partitions   71

House of Boards to build   97

Store House to build, that shall be drift proof   98

Permanent Back-tilts   100

---

1 Entry incorrectly paginated in original. Should read page 60.

---

2 This entry is missing from the table of contents but occurs in the manuscript.

3 The final entry in *Additions*, but not listed in the original table of contents.

ing a wide variety of supplies that Cartwright needed to operate a multi-functional merchant station in late eighteenth-century Labrador. Items such as "Bells, Hawking, Dinner, for Rooms, Cranks for," "Ephemeris," "Lanterns of Glass, Tin, Wire," "Flints, for Guns," "Earthen-ware of sorts," and "Ising-glass" are a small selection of the range of everyday items that Cartwright considered necessary for Labrador life. Many items are obscure and perhaps of lost meaning, while arcana such as "Antinomy, crude," "Nux vomica," "Pansions," and "Udders pickled" each opens up its own area of material culture research. (My notes in square brackets next to some of these entries give a definition wherever possible.) There is a clear connection between "Index for Goods" and Cartwright's *Journal* for many items in the "Index" are mentioned in the latter. Although undated, the "Index" may have been compiled before or during Cartwright's years in Labrador when it had immediate relevance and forgotten items had far-reaching consequences.

b) The second manuscript in this category is the essay entitled "Memorandum for w." Evidently written for a specific individual, this may be Cartwright's draft copy. It outlines the skills and knowledge needed by a merchant trader, particularly with respect to dealing with co-workers and Aboriginal trade partners. Undated and unsigned but in Cartwright's handwriting, it was originally found filed with two letters from 1914 that refer to its loan to Dr Wilfred Grenfell, who appears to have reviewed it as part of his own ongoing work in Labrador.[10] An 1818 letter from the Hudson's Bay Company to Cartwright in the correspondence suggests that the memorandum may have been written for a prospective HBC employee named W. Blagg. The "Memorandum" is also mentioned in an early twentieth-century essay on

Fig. 5   Memorial to George and John Cartwright, in Cartwright, Labrador. Dr Wilfred Grenfell examined this monument in 1893, noting that lichen growth had pried apart the slabs and that he could see inside "a mighty demijohn of rum" (Buckle 2003, 9; see also endnote 26). Photo by M. Stopp 1991.

the Cartwright family by a Mrs George Cartwright, who quoted from it extensively and concluded that "Labrador was evidently no sofa existence" (Cartwright 1909, 128).

2)   A second body of documents is the correspondence. Eight letters written by George Cartwright in 1771 to family members replicate information found in the *Journal* but also contain new details on a variety of subjects of interest to northern researchers, such as the long-gone southern Labrador caribou herd and an early description (possibly the earliest) of an Inuit snow house and a soapstone lamp. A 1774 letter from Daniel Sutton to Cartwright concerning the death by smallpox of the Inuit boy Noozelliack relates to entries in the *Journal* of 15 December 1773. Another letter of 1774 from John Cartwright to his sister Catherine (Kitty) mentions their brother George, and two further letters dated 1818 also briefly mention him. Six letters date to the late nineteenth and early twentieth

Table 4: Listing of Cartwright's Unnumbered Loose Papers

Description	No. pages
"Provisions for one Man for Twelve Months"	1
"Furniture & other necessaries for each Tilt"	1
"Additional necessaries for each resident Tilt"	1
Sheet ripped from narrow notebook with a list of men and women for a station in Labrador	1
Sheet ripped from narrow notebook with list of trap quantities, costs, and locations in southern Labrador	1
"Expenses attending the Publication of my Journal"	2
"A Sketch of a Bay in Labrador copied from one drawn by An Eskimeau Indian" (appears to be coastline between Makkovik Bay and Kaipokok Bay, with small squares possibly showing locations of Inuit sod houses)	1
Untitled list of employees and their wages – probably written during Cartwright's early years in Labrador when he had a surgeon on staff	1
A table showing the number of crew at Seal Island, Hawk Island, Venison Harbour, and "At Home," probably written between 1770 and 1774, before Cartwright began the Sandwich Bay operations	1
"Plan of Buildings for Labrador"	8

centuries and concern the 1861 placement of a memorial to George and John Cartwright in Labrador by their niece, Frances Dorothy Cartwright. This memorial can still be seen in the cemetery of the community of Cartwright (figure 5). In the style of a sarcophagus, it is of white marble with a movable lid. Oral accounts suggest that it once contained items belonging to Cartwright, which have long since disappeared.

3)   Finally, the new papers finish with a number of loose and unrelated sheets on various themes. As listed in table 4, there are notes on the small winter shanties known as tilts, on trapping, and on provisions for one man for a year. To give them context, a few of the loose papers have been incorporated into the discussions that

follow on trapping and house construction, specifically the building plans and the information on trapping tilts.

Other items of particular interest in this category follow the correspondence. They include Cartwright's list of costs incurred in publishing the *Journal*, revealing the £7-7s. paid to W. Hilton on 20 May 1791 for painting the oil portrait. A second unique document is a sketch map of the Makkovik area that identifies what are probably Inuit settlement locations, with a notation in Cartwright's hand explaining that it was copied from a map drawn by an Inuit. Also of historical interest is a list of costs of provisioning one man for a year, and two lists of employees; the two latter lists may have been drawn up in the early 1770s, before Cartwright moved to Sandwich Bay, because they refer to small crews at Seal Island, Venison Harbour, and Hawk Island, and "At Home," which in this case would be either Lodge Bay or Stage Cove. A final list includes wages for crew including coopers, a smith, furriers, "youngsters" (inexperienced apprentices), and women, and the costs of shipping.[11]

CHAPTER FIVE

Historical Relevance of the New Cartwright Papers

Documents such as the new Cartwright papers gain relevance when their contents can be cross-checked against known information, and when that information helps us understand a time or an event in a new or better way. Because much of the material here was written after Cartwright lived in Labrador, it was important to consider its relevance for historic study. In other words, I had to consider the possibility that Cartwright's notes might be fireside musings that never saw field application and consequently held little of historical value. His plans of house structures, for instance, raised the very pertinent question of whether such houses were ever built in Labrador, and whether the other instructions were representative of Labrador life. Did he build the studded house and the seal-oil skimming facility described in *Additions*, and had he hunted caribou using deer fences, or were these all devised and improvised to occupy himself in his golden years as an armchair adventurer?

It can be said that at their broadest the Cartwright papers correctly reveal many aspects of that time and place. Their focus on sealing, fishing, hunting, and trapping fully conveys the central position of these pursuits for early Labrador merchants and settlers, where every success depended on well-functioning traps, efficient hunting and sealing technologies, and health and safety through weatherproof buildings and reliable provisioning of food. Cartwright's lists of goods represent the material means of achieving these ends, while his building instructions reflect techniques and ideas about space that ensured a measure of comfort. Taken together, the papers give some understanding of the wide range of technical knowledge needed by an early Labrador resident and the many material items that facilitated this resource-based economy. Above all, the information gives us a humble respect for a way of life that was one of nearly continuous labour under conditions that were rarely favourable – a way of life that continued nearly unchanged in Labrador until the twentieth century.

By identifying links between the papers and the *Journal*, and with other historic documents, it is evident that Cartwright's information fits into known practices of eighteenth-century Labrador and that he was writing from knowledge based on experience when he compiled these papers. There are, nonetheless, entries in *Additions* that cannot be pinned down and that appear to have little connection with Cartwright's Labrador years. These include instructions on how to train hawks (4, 6). We know that he enjoyed falconry as a pastime in his latter years, as barracks master of the Nottingham militia, but there is no record in the *Journal* that he brought hawks to Labrador, although the "Index of Goods" lists hawking paraphernalia such as hawk bells, whistles, hoods, and dogskins. Brief instructions for a bridge "for an army to pass over" (9) may have been inspired by the small British garrison posted at York Fort in the early 1770s, or perhaps by the need for military protection on a coast frequented by American privateers – but again there seems to be no evidence that this was ever considered for Labrador.

Other entries in *Additions* seem fanciful at first reading but are in fact founded upon real experiences. Directions for the seemingly implausible "sailing sled" (81, 101) are based on a craft that Cartwright enjoyed during his time in Labrador: "We set the sail in the boat [a salmon punt], to assist the dogs, and I sailed down in her the greatest part of the way. The dogs sometimes found it difficult to keep ahead of the sled" (c.25.3.71).

Similarly, the plan for an elaborate storehouse for a bull, cows, horses, poultry, and storage of hay, hops, and maize is not far-fetched. Over the years Cartwright brought a wide range of livestock to Labrador (albeit with a high winter mortality rate) and experimented with planting oats, rye, barley, wheat, maize, cress, lettuces, cucumbers, and fennel (Stopp 2001).

The main themes of the new Cartwright papers – sealing, hunting, trapping, salmon and cod fishing, building construction, and the Aboriginal trade – are developed further in the following sections to show, firstly, where Cartwright's information intersects with known, historical facts. Each section also includes a brief history and description of the subject area in order to provide an historical context for Labrador matters still relatively inaccessible to a general readership.

SEALING

As early as the 1600s French entrepreneurs recognized the immense value of the Gulf of St Lawrence seal fishery to European markets. By the mid-1700s, French sealing posts dotted the Gulf coast and the Strait of Belle Isle as far north as Chateau Bay. Sealskin was used for muffs, boots, shoes, and jackets. Seal oil, along with whale and cod oils, was a commodity of considerable value, traditionally used in leather and fabric preparations. Demand rose by the 1770s throughout Europe with breakthroughs in public and private lighting technologies. During the early years of the Industrial Revolution and the mass production of machinery, seal oil became increasingly important as a lubricant (Ryan 1994). The oil was obtained by cutting seal blubber, sometimes called "rand fat" or "ran fat" by Cartwright, into strips and liquefying it either in vats over a fire or leaving it in vats in the sun (cf. Hallock's observations 1861, 594; figure 6).

Sealing was a costly undertaking and required a considerable outlay of capital for nets, winter shelter and food for crews, fuel, equipment for the rendering process, and the means of bringing the rendered oil to market at a profit. Unlike in Newfoundland, where the early seal fishery was conducted mainly from vessels in ice floes, the early Labrador seal fishery was net based and carried out from shore with some use of boats to position and clear nets. Twice yearly, in late autumn and early spring, large numbers of harp seals (*Phoca groenlandica*) migrate along the coastal periphery, funnelled northwards through the Strait of Belle Isle during the spring migration to reach the pack ice and eventually Arctic waters, returning southwards in late autumn to the Gulf of St Lawrence. To catch them, well anchored shoal nets were strategically positioned at natural capture zones such as narrows between two shorelines or at the mouths of small bays and coves. Shoal nets were usually forty fathoms long (a fathom being about six feet or 1.8 metres) and two fathoms deep and were extended from an onshore capstan or other landfast point. They hung along the sea bed using a system of lines, floats, and killicks (handmade anchors of wood and rocks). The nets were kept in an upright position to catch the seals as they ranged along the bottom after fish. Stopper nets served the same purpose as shoal nets but were custom made to fit a specific location. Their bottoms lay upon the sea bed and their tops at the surface using floats. One end would either be landfast or attached to the head rope of another net

Fig. 6   A Labrador seal vat for rendering blubber into oil, ca. 1917. LAC, PA-175487, photo by Mrs E. Motaling, a missionary assistant to Wilfred Grenfell

moored parallel to shore, while the end closest to shore was attached to capstans and could be raised and lowered. When pulled tight, several of these nets at certain distances from one another formed pounds to enclose seals (Cartwright 1792, glossary). The earliest description of such a netting approach was recorded by Joseph Banks during his 1766 summer in Chateau Bay:

> If they have a narrow Streight between two Islands or an Island & the main which is much the most convenient ... Situation it is Crossd by a number of nets the Last of which only is Drawn tight the rest remaining Close to the Bottom of the Water the Seals who Come in Shoals

finding themselves Stopd by the tight net Crowd to it trying to find some way of getting on in the mean time the fishermen Draw tight the second net by which they are inclosd in a pound the Second Shoal of Seals are stopd by the second net & securd by the third & so they Proceed till they have filld all their nets. (Lysaght 1971, 145)

Southern Labrador's contribution to sealing seems to have been the seal-netting frame, which took netting technology one step further with the use of a blocking net that ran parallel to shore at the seaward ends of the pound nets to prevent seals from escaping (cf. Mosell 1923; Candow 1989). The blocking net, described and used by Cartwright, may have been one of his innovations, since Banks's description does not mention it. The use of the frame dates to at least the mid-eighteenth century (Mosdell 1923) and was described a century later by a visitor to Labrador: "A solid frame of nets is fixed in a convenient place, into which the seals enter – get entangled among the nets – drown, and are taken ashore in boats. This is the process of the seal fishery, as practised in and above the Straits of Belle Isle" (Robertson 1843, 36).

The Strait of Belle Isle was more favourable for sealing than the coast between Cape Charles and Sandwich Bay. Acting as a natural funnel, it contained the seal pack and served as a well-defined capture zone. Early eighteenth-century French merchants and nineteenth-century Jerseyian and British merchants conducted reasonably successful seal fisheries for many years in the Straits, while their colleagues to the north faced a variety of challenges. The coast between Cape Charles and Sandwich Bay lies open to the forces of the Labrador Sea, where a combination of inclement weather and ice movement could quickly alter the course of the two-week seal harvest. In some years Cartwright experienced deceptively mild temperatures in late November when the seals arrived, followed by a sudden freeze that damaged nets and brought an immediate end to the sealing season. In other years strong winds drove immense ice floes against shore, blocking the approach of seals or dispersing ice and seals far offshore. Perhaps the only advantage of this coast for sealing was the many islands close to shore that permitted a shore-based net fishery.

By the mid-1800s many of the prime sealing locations of a century earlier had evolved into seasonal communities from which both the seal and the summer cod fisheries were carried out. "A good sealing post is ranked as of the most valuable species of property, and is transmitted from one family to another," observed Reverend Ephraim Tucker (1839, 107) following a voyage along the gulf shore and the Strait of Belle Isle.[12] Most of the fishing communities of south-central Labrador had their beginnings as sealing stations. In the St Lewis Inlet area, Cape Charles, Battle Islands, Great Caribou Island, and Spear Harbour all began as sealing stations in the 1770s, as did Granby Island and the Fishing Islands at the mouth of Gilbert Bay, Square Islands at the mouth of St Michael's Bay, Venison Islands, and the entire Seal Islands Harbour area. Cape North and the islands and coves at the mouth of Sandwich Bay had similar beginnings (figure 7). Today, many of these coastal communities are deserted as a result of the cod moratorium, and the bulk of the population lives in inner coastal towns that began as winter communities for trapping or logging, such as Lodge Bay, Mary's Harbour, St Lewis, Port Hope Simpson, Charlottetown, and Cartwright (the exception being Black Tickle, a coastal community that is inhabited year round).

Cartwright's many notations on sealing in *Additions* were inspired by the challenges of trying to capture a continuously moving species under equally shifting ice and weather conditions. His notes relate to innovation and improvement of the production process and the housing of sealing crews, and they tie in well with information on sealing found in the *Journal*. Sealing was a persistently difficult enterprise that may have realized Cartwright negligible profits over the years. Although the autumn seal fisheries of 1770 and 1771 brought excellent harvests, those of 1774, 1775, 1777, 1778, and 1785 were all poor, as were the spring seal fisheries of 1776 and 1777. (Missing years are due to a number of factors. During some seasons Cartwright did not record a seal fishery because he was in England. During others, seal nets were lost at the beginning of the season due to ice or wind, effectively ending that fishery. In 1773 he lost his entire seal and salmon fishery to the takeover by Noble and Pinson who recorded a profit of £16,591 from sealing at Cape Charles the next season [Mosdell 1923, 19].)

Cartwright's instructions on sealing in *Additions* were founded on the premise that a well-planned operation increased the probability of success even though the movement of seals could not be controlled. He was a vigilant and attentive sealing-master, paying close attention to the strategic placement of nets to develop an optimal seal capture zone, and to

Fig. 7   *Above*: A typical Labrador wharf, Venison Tickle, 1886. LAC, e003894434, Pinkney collection; *right*: Fishing schooners at Cape Charles in 1909. LAC, C-065061, photo by W. Notman 1909

the oil-rendering process to produce high-quality oil. In a letter to Lord Dartmouth in 1773, thirty-nine years before writing *Additions*, Cartwright referred to his efforts at innovation and improvements in seal harvesting and processing: "[I have] by great Industry and observation, and by many chargeable Experiments, invented several Considerable Improvements in the Mode of carrying on the Seal Fishery, whereby it will be managed with Greater ease, the Oil will be much more pure, and produce a greater profit, by at least 20 per Cent" (CO 191/31, 12v). In *Additions* (22) he mentions further sealing innovations, referring to square and bent pumps that he invented for a sealing shallop.

The instructions recorded in *Additions* illustrate his broad knowledge of sealing. His first entry on the subject, "General Directions for a Sealing-Post" (*Additions*, 12–14), is emphatic on the importance of proper gear and choosing a sensible location: "The proper places to catch [seals] are either

on the South side of a Bay and near to the mouth of it, or between an Island and the Continent upon a straight shore ... The greatest errors which I have observed in catching Seals, and which were committed at every Post that I have yet seen, were the grudging the expense of a sufficient quantity of Twine and Cordage, both in the number and depth of the Nets."

Cartwright's plans, instructions, and calculations for sealing (listed in table 3) all point to standardization as a way of improving production. The "Plan of a Dwelling-house for a Sealing Crew" (*Additions*, 19), for instance, shows a building of 68 by 26 feet modelled after dwellings built for crews in 1774 at Stage Cove in St Lewis Inlet (70 by 25 feet), and in 1782–83 at Paradise River in Sandwich Bay (60 by 25 feet) (C.27.8.74; C.11.9.83).

The plan in *Additions* for a sealing post at Venison Island is intriguing and appears to be based on Cartwright's knowledge of that place. His intent here was to provide direction for a net fishery in a tickle, which is a narrow passage of water between two shores. Venison Island has a long history of use as a sealing post. As with the Cape Charles Islands, Battle Harbour, Frenchman's Run, and Gready Islands, the position of Venison Island as a small, elongated piece of land alongside a larger land mass created a natural capture zone for seals. Cartwright was probably the first to place a crew here in the early 1770s, as suggested by one of the employee lists in the loose papers that shows a sealing crew of fourteen. In 1785, "a new adventurer on the coast" by the name of Hyde, working for Thomas Slade, built what Cartwright described as "a very good sealing-house" at Venison Harbour (C.5.8.86; C.6.8.86). Preliminary archaeological investigation there in 1991 yielded early nineteenth-century ceramics, but continuous settlement has removed any surface traces of the earliest buildings (Stopp 1995).

A plan in *Additions* (28, 29) for setting sealing nets at Cape Charles describes another type of shore-based seal fishery. In this case Cartwright used Cape Charles as a model for sealing in a wide tickle or along a straight, open shore, which required somewhat different net placements than in a narrow tickle. Cartwright's first sealing establishment was in fact at Cape Charles in 1770 and 1771, and his instructions are again based on direct experience at that location. In December 1770, for instance, he had twelve shoal nets at Cape Charles of forty fathoms by two, and three stopper nets of 130 fathoms by six. The latter were moored fast to today's Tilcey Island (formerly White Fox Island)[13] and the other end to capstans on today's Wall Island (formerly Seal Island), which allowed sealers to raise and lower the nets. Arranged about forty yards apart, they formed two pounds. A net was also stretched from Wall Island to the mainland to prevent seals from escaping out to sea (C.3.12.70). The plan shown in *Additions* follows the essentials of the one described in the *Journal*, whereby capstans were to be placed on Wall Island and a blocker net stretched to the mainland, with the chief difference being that this new plan was for a larger installation requiring more nets. Apparently Cartwright thought that the harvest in this location could be maximized if one were willing to risk greater investments.

Cartwright's plans in *Additions* for oil and skimming houses, oil-rendering vats, a seal-skinning house, a salt house, and a net house (listed in table 3) are clearly based on his Labrador practices. Although the *Journal* contains none of the detailed information on sealing infrastructure found in *Additions*, occasional references affirm that it was part of Labrador life: "I gave Attuiock five harp skins to cover his kayak; and he carried three larch planks to Seal Island. Those, and the other planks, which

I have sent down at different times before, are intended to build vatts [*sic*] for the seals' oil, when it is melted out in the spring" (C.5.3.71).

The information on sealing in these new Cartwright papers covers many aspects of that industry and counts as some of the earliest first-hand material on the subject. While not all-inclusive (for instance, nothing is recorded about carcass-processing or barrel-making), the information on materials and techniques represents eighteenth-century approaches to a nearly forgotten industry.

HUNTING

If sealing held Cartwright's attention in the autumn and spring, hunting preoccupied him throughout the year. As already noted, fresh meat secured workers' health and productivity and was more economical than foods brought from Britain. The *Journal* contains nearly daily accounts of hunting and trapping exploits. The new papers, in contrast, contain surprisingly little on the subject, and one can only speculate that it featured elsewhere, perhaps in the lost manuscript *The Labrador Companion*. The exception is several entries on capturing caribou using deer fences.

Cartwright appreciated everything about caribou; antlers were a proper huntsman's trophy, the skins had trade value, and venison was his favourite food. He even had a devoted pet caribou calf that for a time accompanied him on walks and slept on the floor by his bed (C.14.8.79). Any sighting or slotting of caribou was judiciously recorded in the *Journal* with palpable pleasure. In the 1700s caribou appeared with regularity along the southern coast, though for the last century there have been few and certainly no regular herd movements. The herd that Cartwright knew appears to have been over-hunted in the intervening years or has shifted its migration pattern. A vivid description of its movements is found in the Cartwright correspondence in a letter to his brother John, written at Ranger Lodge on 20 September 1771.

Several entries in *Additions* contain detailed information on capturing caribou using deer fences. Cartwright first saw these in 1768, during the expedition with his brother into the interior of Newfoundland to find the Beothuk Indians. Following the course of the Exploits River they encountered a stretch of over fifteen kilometres of fence works. These had been constructed by felling trees so that the trunks did not break apart completely but fell upon one another to form a continuous barrier high enough to prevent caribou from jumping over. Where trees were scarce, barriers had been made by substituting trees with sewel sticks, which were slender poles stuck into the ground at intervals, with each pole set at an angle. Hanging from the pole's tip was a sewel, consisting of narrow strips of birchbark tied to form tassels that moved with the slightest wind and made a rasping sound. Caribou would shy from these and be effectively directed into a pound or to a strategic point where they could be killed (Marshall 1996, 276, 328). Similar techniques were used elsewhere in the eastern Canadian woodlands by Amerindian groups including the Iroquois (Tooker 1967, 65; LAC "Champlain") and the Innu of Labrador (Leacock and Rothschild 1994, 115; Pasteen n.d.).

Upon arriving in Labrador in 1770, Cartwright was quick to implement the use of deer fences. During his first winter he had workers "cutting sewel-sticks" and building deer fences throughout Charles Harbour and on Eyre Island (today's Assizes Island) (C.29.3.71; C.3–10.4.71). These sewels were made of boughs, twine, and eider feathers, and one entry refers to using wire for a "deer net" (ibid; C.4.11.85; C.27.2.86). Making sewels, sewel sticks, and deer fences were activities recorded each winter and spring during his residency at Ranger Lodge (1770–74) and again during his years at Isthmus Bay (1779–86). Whether these fences were an effective means of catching caribou remains debatable. One *Journal* entry from Cartwright's time in Isthmus Bay relates that when caribou were chased into the deer fence, they continued running through and under the sewels (C.28.2.86). In characteristic fashion, he innovated on the Beothuk deer-fence prototype observed on the Exploits River. At Isthmus Bay he recorded, "I set [the crew] to work to erect a pound of my own invention, for catching any number of deer alive" (C.18.10.79). This fence managed to capture two caribou that then escaped by running into and breaking the rails. Although his "invention" is not described, the reference to rails suggests that it may have served as a template for the many entries and plans on deer pounds in the new papers that show elaborate and labour-intensive use of posts, boards, nails, and stakes with fodder, and lack the simplicity of the Beothuk design. Cartwright's intention in recording plans for these fences may have been to provide direction to a future inhabitant of the coast, yet it is doubtful whether any European besides himself ever used deer pounds in Labrador; the proof of their usefulness, however, may lie in the fact that Cartwright constructed them over and over again.

TRAPPING

The fur trade was a sideline of sixteenth-century Basque whalers and also of early Dutch, English, French, and Spanish voyages that sought baleen, whalebone, walrus tusks, seal and whale oil, walrus oil, and of course cod (Barkham 1980; Cell 1969; Turgeon et al. 1992). There is little evidence to suggest that the earliest sealing and fishing stations in Labrador had specialized trapping crews. In the land grant to Sieur de Courtemanche in 1702, the main commercial activities were stated to be trading and sealing (Roy 1940, 16). In the late 1730s the trader Antoine Marsal, who held the land between Cape Charles and Alexis River, also held contractual stipulations for trading along with sealing and the cod fishery but not for trapping (LAC, MG1, f68, 77; PC 1927, 3663; Roy 1940, 52, 74). Trade was considered the main source of furs, while in winter sealing crews occupied their time with casual trapping.[14]

Until Cartwright's time the majority of furs obtained by Europeans in southern Labrador and on the Quebec North Shore were from Innu and Inuit who traded for commodities such as kettles, tools, and other items of iron and brass, as well as for beads and fabric. This contrasts with the Island of Newfoundland where, beginning with John Guy's colony in the early seventeenth century, English settlers trapped their own furs, largely because of the Beothuk's reluctance to trade (Pastore 1987, 50; 1989, 57). In Labrador, however, Innu and Inuit continued to be a key source of furs into the late 1800s. They trapped in traditional land-use areas such as the rivers flowing into the Gulf of St Lawrence, the Churchill River system, and along the many river systems north of Hamilton Inlet.[15]

The earliest recorded evidence of European winter trapping for coastal Labrador, albeit conducted under the auspices of Innu hosts, occurred in Hamilton Inlet in 1743, when Louis Fornel left two men to overwinter there in the care of a group of Innu. The men's job was to identify sealing places, learn the Innu language, and explore this enormous watershed. (Fornel believed that one of the rivers reached Hudson's Bay.) The following spring Fornel's vessel failed to reach the Inlet, and his men returned to Mecatina along an Innu overland route, bringing with them a large supply of marten pelts (White 1926, 6: PC 1927, 3336, 3669; Roy 1942, 118, 204; Anick 1976, 620).

Following the 1774 Quebec Act, which allowed ownership of fishing premises, British merchants were quick to establish themselves on a year-round basis. Cartwright and other merchants had winter trapping crews in place along established traplines that extended inland from the heads of all the bays and inlets between Chateau Bay and Sandwich Bay. What little we know of eighteenth-century trapping in southern Labrador comes from Cartwright's *Journal* and from the new papers. Even recent ethnohistoric sources provide only sketchy information on aspects of winter activities such as trapping. One exception is W.H.A. Davies's detailed descriptions of mid-nineteenth century life in Hamilton Inlet. Davies's information is worth quoting here as it sheds light on information left by Cartwright in the new papers:

> Generally about the middle of September, the men are sent into winter quarters; that is, they are sent in parties of two each, up the different rivers, to pass the winter in trapping martens and other animals; they live in small huts, warmed by a stove; their work consists in visiting their traps, keeping them free from snow, and in hunting for a part of their subsistence. Their traps are either steel ones, or made of wood, technically called "dead falls" – these latter traps are constructed in such a manner, that the animal on taking the bait, pulls down a heavy piece of wood that crushes him. The traps for martens are placed along a blazed path (called a "cat path") leading into the interior, and varying in length from one to three days' walk ... the traps for foxes are placed along the borders of the rivers or bays. (Davies 1843, 87)

Cartwright learned to trap after coming to Labrador, probably from crew members who were old Labrador hands. His techniques were an amalgam of European and Aboriginal ones, using commercial steel traps alongside constructed wooden deadfalls. Typically, he was quick to improve on his newly acquired knowledge. While still a fledgling trapper he recorded, "I walked Island Brook Path, where I found all the deathfalls frozen, and two

of them broken. At night I planned new ones, and made models of them"
(C.10.12.70). He continued to improve on trapping techniques throughout
his time in Labrador. In 1777 he asked a veteran trapper named Wrixon,
a planter for the merchant Jeremiah Coghlan, to accompany him along
a Sandwich Bay trapline for advice on the art of tailing (setting) traps
(C.31.10.77). Not long thereafter, however, Cartwright was amending what
he had learned, "I do not approve of Wrixon's way of covering foxtraps with
canvas for on smelling it, they pull it off; I will therefore practice it no
longer" (C.11.12.77).

During his first winter of 1770–71, he established traplines that
extended into the hinterland from Ranger Lodge. In later years he positioned trap crews at greater distances from his main residences, where
they remained until spring and the start of the sealing season. Trappers
worked their lines all winter, sometimes making short returns to the
merchant station for Christmas if it was within reasonable walking distance. They subsisted in part on flour, sugar, lard, and tea, as well as par-
tridges, rabbit, and other eatable meats caught on the trail. Fur-bearers
such as fox, otter, wolf, and muskrat were usually not considered good
eating, although in the spring of 1779, when Cartwright's men were low
on food and close to mutiny over small rations of salted pork, he managed
to convince them that white fox was tasty by eating some of it himself with
gusto and making out that it was superior to hare. The crew quickly followed suit, thereby solving the problem of meat supply (C.1.6.79; Lysaght
1971, 77). For winter shelter, trappers lived in small, rudimentary log
cabins called "tilts," spaced the average distance covered on a winter's day,
generally about four hours' walk.

The way of trapping described by Cartwright continued essentially
unchanged among the Settler population until the introduction of the
snowmobile in the mid-1900s, and traplines are still passed from one
generation to the next. Although the number of traps that hang unused in
storage sheds has increased over the past fifty years, trapping continues to
form a strong part of the Labrador identity (PANL – Jerrett; Tanner 1947;
Goudie 1973; Zimmerly 1975; Goudie 1991; Borlase 1994; Buckle 2003;
Cockerill 2004; see also many editions of *Them Days* magazine).

Trapping fur-bearers such as beaver, mink, marten, fox, wolverine,
lynx, and wolf, snaring rabbits, and shooting the occasional black bear

and polar bear constituted a lucrative part of Cartwright's enterprise, especially during his years in Sandwich Bay. His portrait by W. Hilton is fittingly entitled "Captain Cartwright among His Fox Traps" and shows him against a barren winter landscape with a trapped fox slung over his shoulder and another in the background caught in a leg-hold trap. His many references to the habits and capture of beavers, both in the *Journal* and *Additions*, were undoubtedly prompted by the importance of this fur-bearer to the Hudson's Bay Company and the North American fur trade in general and ultimately to garnering interest in his enterprise among investors. Ever innovative, he introduced a plan for beaver and fox farming (*Additions*, 56, 61) by giving directions on how to capture these creatures live and maintain them.

Entries on trapping in *Additions* present the specialized techniques and language of trapping first encountered in the *Journal* and also given in Davies's quote above. The route of the trapline was called a "path," while a "cat-path" was the marten trapline. Paths were given names that linked them with recognized toponyms; those paths that led from Ranger Lodge, for instance, were called Nescaupick Path, Hare Hill Path, Prospect Hill Path, Island Brook Path, and Snug Pond Path. Cartwright's traplines followed waterways for capturing marten, muskrat, beaver, and otter, but also the footpaths of larger carnivores such as lynx in wooded areas and on the open hinterland barrens. Trap types included springed metal leg-hold and head traps that were tailed for otter, foxes, and wolves and protected from heavy snowfall by a hut of boughs, or a "cat house" for marten. Deadfall traps, which Cartwright referred to as "deathfall" traps, were gravity based, whereby a bait stick tripped by an animal caused a carefully aligned series of logs to both capture and crush it.[16] Deadfalls could be constructed for animals of any size but were mainly used for marten. Snares, or slips, involved a wire tightening around the animal's neck or leg as it stepped into the snare and attempted to move away. They were used for rabbits but also for larger mammals such as fox and even caribou and bear. Beavers were captured with nets or by breaking into their lodges from above, a technique that Cartwright learned from Innu. Drags for scenting an area to attract game and the use of poison and deviously placed hooks and baits were also part of the trapping repertoire; Cartwright mentions using a mix of Cheshire cheese and honey to bait fox traps in Sandwich Bay, and "poison balls" on Huntingdon Island to catch fox (C.10.12.77; C.29.12.77), both ideas that reappear in *Additions*.

Additions also contains early information on building a trapping tilt and constructing a marten path from Deer Harbour in St Lewis Inlet to Hawke River. Cartwright's notes give some idea of the knowledge, decision-making, and capital investment involved in successful trapping ventures, including planning where crews would be stationed for the winter, the placement of tilts along a trapline, and the crucial timing of supply vessels that brought winter supplies for the trapline. The number of traps needed at any location also had to be well planned. Such a list, jotted on one of the loose papers and probably written before moving to Sandwich Bay in 1775, gave the number of otter traps for St Lewis Bay as thirty-six, with forty for Alexis Bay, thirty-six for Gilbert's Neck, forty for St Michael's Bay, thirty-six for Hawke Bay, and twenty for Caplin Bay, with a further eighty-four head traps for each location, running to a total cost of £111 8s. for traps.

The loose papers also include two lists of items needed at each main winter tilt. They suggest a practical but well-supplied trapping operation and are as follows:

Furniture & other necessaries for each Tilt
1 Stove & a set of Funnelling & a Potcrane. 3-2½ wide x 2-9 high
1 Window of 4 Squares of Glass made to be double glazed
1 Table 3 x 2 feet with two Drawers
2 Iron Pots, 1 Tin Kettle, 2 d° sauce-pans, 2 d° Pudding-pans
2 Block-tin soup & 2 d° flat-plates, 2 Quart pots
2 Hatchets, 1 2 in.[?]
1 small hammer, 1 Drawing-knife, 4 Files, 1 Razor stone
1 Charwood Forest stone, 6 small skinning knives, 4 larger d°, 2 Butcher's d°, 1 Steel d°.
8 Marten traps for each mile of trail, 6 Wolf d°
50 lb shallops-nails, 28 lb /2 nails, 100 Marten-boards, 4 Wolf d°, 4 Wolvering d°, 2 Lynx d°, 2 Fox d° 6 Deer-slips, 4 lb gunpowder, 50 Flints, 1 Bag of shot No. 4 Patent, 10 lb Mould-shot.
2 Doors 4 feet by 2-6 with a board 1-6 at bottom for the door to shut upon. Hinges for doors & shutter for window – 1 Canoe (where necessary), Oakum.

4 Gimblets, 400 Brads, 1 Brad-awl. 2 Bedsteads 6 x 4-6, with sacking-bottoms, 10 1 in, 10 1 ¾ & 10 ½ in boards.

1 morticing Chizzel & mallet (to be made there) 3 Featherbags, Fixed & hanging shelves, 1 Tierce of Meat-pickle.

3 empty Hogsheads, some small Kegs, 2 Beer-kegs [ink blot] Gallons each (one pint of Treacle). A Cellar, to be made when convenient, Pitch, Tar, Brown paper.

1 Lamp, Cotton wick, 1 Candlestick, 1 large snuffer

1 Handsaw

listing for Door-cases, a Deer-skin Window curtain, 1 Iron ladle of ½ pint. 1 Tinder box & steel

12 ¼ lb Dipped Candles 8 to lb.

2 splitting knives, 2 Busks for each Furr-board, 1 Jack, & 1 [ink blot]

1 Smoother Plane, 2 oz. [?] Vivium

Provisions

1 B^l. [barrel] Meat, 1 ½ C^{wt}. [hundredweight] Bread, 1 C^{wt}. Flour, 6 Gal. Treacle, 4 d° Oil.

4 d° Vinegar, 8 Gal. Pease.

Additional necessaries for each resident Tilt

2 Guns, 4 lb Gun-powder, 2 Bags No. 1 shot, 10 lb Mould-shot, 50 Flints, 3 or 4 doz. Otter Traps, 1 Bear d°.

1 small Skiff, ½ Cwt Shallops-nails, 10 lb ½ spikes.

1 large Hammer, 1 Grind-stone & winch, 1 small Bread-box, 2 Knapsacks, 2 pr. slings, 2 Nescaupic sleds 6 x 1 foot.

2 Shovels, 2 Spades, 1 Pickaxe, 2 small hatchets, 1 spare large one, 2 Flour-sacks for Sleds, 32 Powder-flasks, 2 double shot-makes, 2 Turn-screws, 1 Pint scenting for Traps, 2 Phyals with ground stopples in tin cases for carrying scenting, 2 Bait-bags, 2 Bullet-moulds, account Book, Memorandum d°., 1 quire of paper, 25 quills, 1 Pen Knife, 1 stick Indian ink, 10 ¼ lb Candles, 8 to the pound dipped (for the first year),

2 ½ Peter lines for otter-traps, 2 Beaver-net, 40 x 8 feet, mesh 3 ½ in square, 5 lb Lead, 1 Skein of Salmon twine. Thief nets for small Brooks for Otters; twine, Shoal-net, mesh 2 in. square, diameter 18 inches. 2 Gun-caps of Dogs skin, 30 Otter Boards, 2 Busks for each, 1 Pike-hook & line.

FISHING

Salmon and cod supported an important commercial fishery in southern Labrador between 1700 and the 1992 cod moratorium. In the 1700s a migratory French/Breton, ship-based cod fishery operated for a three-month period each year, while Quebec-based merchants had a few permanent sealing/fishing posts in the Strait of Belle Isle and on Newfoundland's French Shore.

After 1763 the English merchants first based in northern Newfoundland began to expand their cod and sealing operations into harbours along the Labrador coast such as Forteau and Temple Bay. By the late 1700s the Straits and all the way north to Groswater Bay were being exploited by an ever-increasing number of fishing vessels arriving each summer from Newfoundland, the Atlantic colonies, Quebec, France, and New England, in addition to those vessels belonging to land-based merchants such as Cartwright.

Both Cartwright's *Journal* and his later papers are surprisingly uninformative on the cod fishery, with only occasional references to bait fish or the construction of a fishing stage. (It is possible that *The Labrador Companion* contained more information on this industry.) His focus as a chronicler was largely taken up by hunting and trapping, although the cod fishery seems to have been a constant but uneventful activity for him. His main cod-fishing stages were situated at Stage Cove and Great Caribou Island in St Lewis Inlet, and on Great Island at the mouth of Sandwich Bay, while his most valuable salmon-fishing premises were the rivers in Sandwich Bay. During his first five years in Labrador, he participated little in the fishery, admitting in a letter to Lord Dartmouth in 1775 that he "was a great stickler against the Cod fishery" (LAC, Dartmouth fonds, 2460–4), probably because it required a large capital outlay in large crews and many vessels, and perhaps because it meant competing against large firms such as Noble and Pinson. In the same letter, however, he acknowledged the value of having larger crews over the summer because some of them would remain for the winter seal fishery and trapping.

As with the cod fishery, the new Cartwright papers contain only passing references to the salmon fishery, despite its importance to Cartwright's business. The commercial salmon fishery season in the late eighteenth century began in June or July in all the major rivers between Blanc Sablon and Sandwich Bay. The most important of these were Forteau River, Pinware River, Temple Brook, St Charles River, Mary's Harbour River, St Lewis River (which Cartwright called the "Colleroon"), the three streams that flow into the head of Alexis Bay, the Sand Hill River, and the rivers that flow into Sandwich Bay (the Eagle, Paradise, and White Bear Rivers) (Taylor 1985). Most of these rivers fuelled a lucrative salmon industry for more than two hundred years, declining in the 1990s around the time of the cod moratorium. The arrival of salmon dovetailed with the end of the spring seal hunt but often coincided with the start of the summer cod fishery. Indeed, between the late 1700s and the mid-1800s many ships that arrived off the coast of Labrador from Newfoundland and the Gulf of St Lawrence region for the cod fishery also caught salmon, with their intake becoming part of the export record of the home port rather than Labrador (ibid.). For a Labrador merchant the advantages of the salmon fishery over cod were that nets could be positioned and fish caught, cut, salted, dried, and stored near an onshore station without requiring the use of costly large vessels for fishing offshore. In addition, crews remained at the station and were thus available to complete other tasks, especially those related to the processing of seal oil and to barrel-making.

For the eighteenth-century merchant in southern Labrador, the salmon fishery figured alongside the seal hunt as one of the year's chief economic activities. Its importance is iterated time and again in Cartwright's writings as well as in Colonial Office papers for Labrador. A successful salmon enterprise required control of the mouth of one or more salmon rivers, knowledge of where to position nets, vigilance in keeping the nets clear, shore-based facilities for drying and salting fish, materials and facilities for making tierces (forty-two gallon barrels) in which to store it, and buildings for storing it before shipping to market. Salmon was largely caught with fixed gill nets and sometimes with weirs. Gill nets were also used by cod fishing vessels to catch salmon offshore (ibid.). Pickled or salted salmon, along with seal oil rendered and stored in barrels, was shipped to England on early summer voyages and provided the capital to pay for the remaining year's supplies.

Cartwright's largest salmon fishery was in Sandwich Bay, at the mouths of the Paradise, White Bear, North, and Eagle Rivers. His main station in 1775 at Paradise River included wharves, a salmon storehouse with a salt room measuring forty-four by twenty-four feet, a cooperage stage, a fishermen's house with a salt room, and a vegetable garden. Eight years later this station had expanded considerably, with a larger vegetable garden and "a dwelling house and store-house in one, sixty feet by twenty-five, and two stories high; a house for the servants, thirty feet by seventeen; three salmon-houses ninety feet by twenty each; and a smith's shop, sixteen feet by twelve" (C.11.9.83).

As with trapping, Cartwright learned about commercial salmon fishing after he arrived in Labrador, probably on the St Charles River, which flowed past his front door at Ranger Lodge. He quickly expanded to include the St Lewis River, the rivers at Port Marnham, and the St Mary's River. During his early time on the St Charles River, he struggled with the proper placement of nets, their quality (some of his were "rotten"), and with different strategies to increase the catch. In *Journal* entries of 1771, he gives the impression of being an experienced old hand at the intricacies of a commercial salmon fishery, yet his knowledge must have been recently acquired. Some of his know-how probably came from Inuit companions. "Shuglawina, whom I found to be a very intelligent man, and possessed of strong natural parts, advised me to make a pound to catch salmon, and shewed me where to place it," he wrote, adding, perhaps in the manner of a man unused to giving credit where credit is due, "I was greatly obliged to him for his information, although it happened to prove unnecessary; for I have one now making, and intend to fix it in the very place which he pointed out" (C.6.7.71). This device was undoubtedly the precursor to three brief references to salmon "racks" in *Additions*, which also contains instructions on drying salmon, preparing salted salmon, and extracting salmon oil.

CONSTRUCTION

Along with instructions on deer fences and traps, the new Cartwright papers contain several notes on the construction of seal-processing installations, tilts, wharves, and dwellings. Cartwright made sketches of some of these structures, and in many cases also gave measurements, instructions on how to build them, and detailed lists of construction materials, costs, and quantities.

Of particular interest is whether the constructions described in the papers resemble or bear any relation to those Cartwright built in Labrador. Here again, the issue of relevance and the practical origin of his later ideas is worth examining in order to assess the historic value of this information. The majority of references to construction in the papers pertain to dwellings, both for working crews and for a merchant. *Additions* contains thirteen entries on house construction that range from how to preserve the bottoms of studs, the quantities of planks and glass needed, how to build a studded house, and how to fix beams and rafters, to building trapping tilts, wharves, and storehouses.

When this information is compared to descriptions in Cartwright's *Journal* of dwellings built in Labrador, it becomes clear that the later plans in *Additions* were for more ambitious buildings. What remains constant, however, is the basic floor plan and layout, which differ substantially from popular house designs of the time and may represent a vernacular type unique to eighteenth-century coastal Labrador. Cartwright's five main residences during his years in Labrador, located at St Charles River, at Stage Cove, at the mouth of the Paradise River, in today's community of Cartwright, and at Isthmus Bay, were each built following a style that may have had its origins in a medieval English architectural type called the hall. This was a rectangle entered near one end, usually the kitchen end, through one of its long sides. Further rooms were added along the long axis, with each room the full width of the structure.

By the seventeenth century the hall form tended to have a minimum length measurement of thirty feet, with a proper chimney rather than a central hearth, and in architectural parlance came to be known as the longhouse. These rectangular structures varied considerably in size, but were often based on widths of roughly fourteen to sixteen feet or twenty to twenty-four feet, with many variations on length depending on the number of rooms along the long axis. Dimensions loosely conformed to the sixteen-foot measure, which was based on factors and multiples of sixteen, or being divisible by four, and was part of an architectural grammar found in eighteenth-century early American buildings.

By the mid-eighteenth century the longhouse design had been effectively replaced by the hall-and-parlour form in England, where the base unit was a square room attached to another that could also be square, but was usually less than square, and where dimensions still followed the sixteen-foot measure. It was this form that transplanted to the early American colonies, and it was built alongside another popular form, the centre hall-type house. The latter was entered in the centre front for

Fig. 8   Wood frame houses, fishing stages, and other outbuildings at Cape Charles, 1909. LAC, C-065060; photo by W. Notman

both small family dwellings and manor houses, where the width of the house was often two rooms deep. Both of these forms came to represent the "traditional" housing along much of the eastern seaboard, including Newfoundland and Labrador (Braun 1962; Whiffen 1960; Glassie 1975; McAleese 1991; Deetz and Deetz 2000; Mellin 2003).

Well-built hall-and-parlour type houses and centre hall plan houses of frame construction began to dot the shores of south-central Labrador by the mid- to late 1800s with the gradual growth of an economically independent resident fishing population (figure 8). The earliest homes along this coast, however, were rudimentary shelters of studs insulated with an outer sod covering. These structures, housing fishing families who were subsistence earners and in debt to merchants, probably found their origin in the sod-house style of their Inuit forebears (figure 9). The sod and wood house consisted of an open room with a hearth or drum-type stove and sometimes simple partitioning. Early urban travellers to southern Labrador were sufficiently struck by life on the coast to record something of these buildings: "The lodging shanties are constructed of spruce poles or

Fig. 9   A sod house near Batteau, 1893. Courtesy of R. Rompkey, Memorial University of Newfoundland, St. John's; photo by E. Curwen

studs, after the fashion of the 'stages'" (Hallock 1861, 595); and "Houses [are] built of logs sawn in two and placed upright, with the rough side out. These form the walls; the roof is of smaller sticks or of poles for rafters, thatched with birch-bark and covered with 'sods.' The seams are caulked with moss, a floor put in; and with a partition or two, and an immense Canadian 'double-stove' in the centre, you have the typical Labrador house ... Some, however, have frame houses, and are roomy and comfortable" (Butler 1878, 4).

Cartwright's buildings were longhouse-type structures inasmuch as they had a basic floor plan that was longer than it was wide: two rooms for daily living (a kitchen and a dining room) and added space for bedrooms and sometimes storage, all the same width and built along a single axis. Their proportions, moreover, fell roughly within the sixteen-foot dimensional module but were rarely a perfect fit. Cartwright's choice of the rectangular structure throughout his years in Labrador is curious, and its exact architectural lineage is uncertain. It may be an anachronism, or was based on institutional buildings such as military barracks, or on earlier merchant buildings along the Labrador coast – or, as has been suggested, on English hunting lodges (McAleese 1991); hence the name Ranger Lodge.

Cartwright's first home in Labrador, Ranger Lodge, was certainly based on an earlier merchant structure. It was actually built as a thirty-seven by fourteen foot workshop by Nicholas Darby, the previous merchant at that place (C.4.8.70). Upon arriving on the coast, Cartwright converted it into his living quarters, making only a few internal renovations, even though Darby's house, of which we have no details, was in perfect condition (it was turned into the storehouse). For reasons unknown, the rectangular form of Darby's storehouse resonated with Cartwright's notion of a merchant's home.

A possible military prototype for the rectangular structure is suggested by archaeological and archival studies of the remains of Sieur de Courtemanche's eighteenth-century stockaded trade post at Brador, where the main dwelling was a single axis structure seventy-five by twenty-eight feet (roughly twenty by eight metres) in size and divided into rooms (Niellon and Lamontagne 1982). The small English fortification of York Fort, built at Chateau Bay in 1766, contained a barracks with a rectangular base form of forty-four by sixteen feet (note the sixteen-foot module; roughly fourteen by five metres), with two large rooms along a single span or long axis (Lysaght 1971, 451).

Despite its puzzling origins, the rectangular base structure remained relatively unchanged for all the dwellings described in the *Journal* and those presented in *Additions*. Four rudimentary sketches of buildings from the loose papers show the same rectangular form applied to a furriers' tilt and crew quarters (figure 10). Although the proportions vary from house to house, in all Cartwright's buildings there is consistency in the layout of the rooms along the long axis, often with the kitchen at one end of the axis, sitting room in the middle, and bedrooms at the other end. We learn from the *Journal*, for instance, that the Ranger Lodge house was divided into three equal parts, with the south part a kitchen, a central dining room, and the north end subdivided longitudinally into two bedrooms, with a loft over the whole (C.4.8.70). The Stage Cove dwelling was seventy by twenty-five feet with a twenty-four by twenty-four foot kitchen, a twenty-four by sixteen foot dining room, six bedrooms, and a small mid-passage, and had only one level (C.27.8.74). The Paradise River

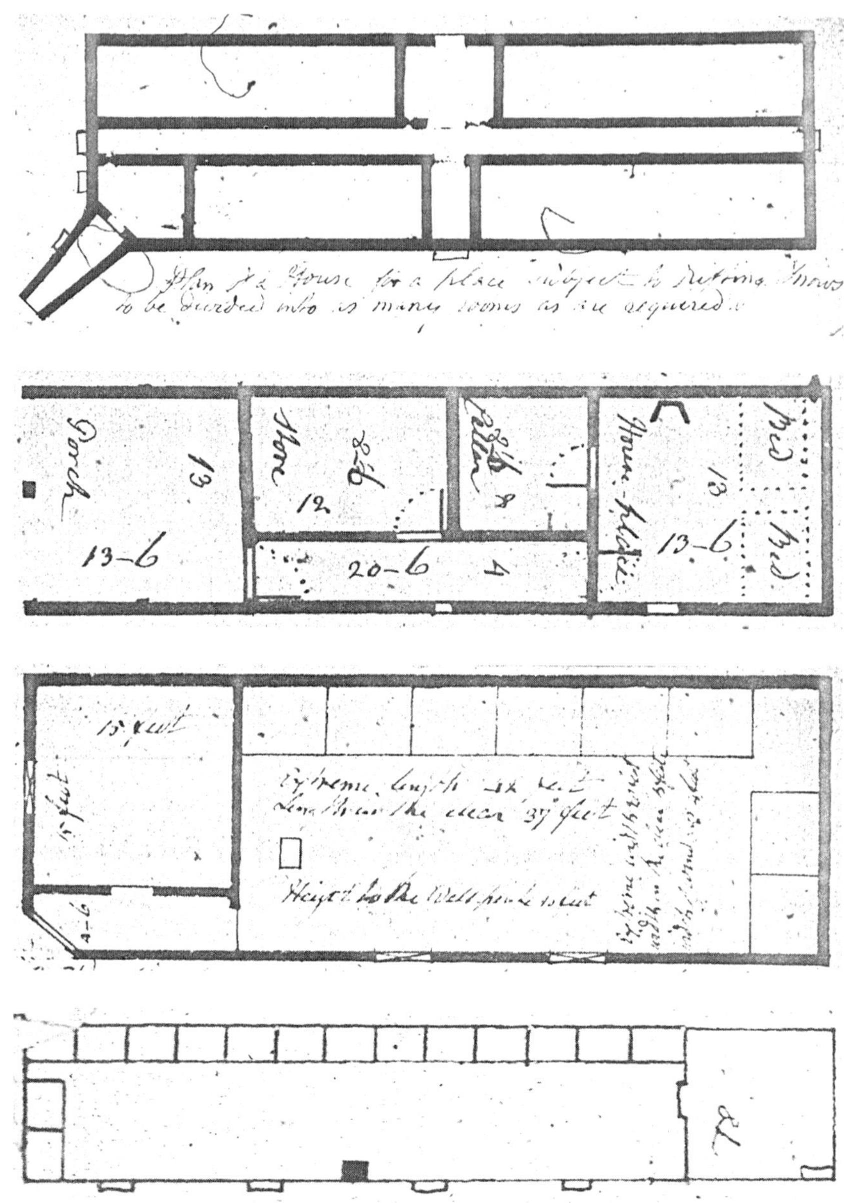

Fig. 10   Four plans for rectangular buildings drawn by Cartwright on loose papers.

establishment consisted of a two-storey dwelling house and storehouse in one, sixty by twenty-five feet. There was also a servants' dwelling house of thirty by seventeen feet, and three salmon houses of ninety by twenty feet, and a smith's shop of sixteen by twelve feet (C.11.9.83).

Caribou Castle, Cartwright's longest place of residence at today's community of Cartwright, is unfortunately hardly described in the *Journal.* Now destroyed by modern-day construction, the remains of this structure were still visible in 1858 when the Reverend William Grey published a series of early sketches of coastal Labrador, noting (but unfortunately not sketching) that "the foundations of [Cartwright's] house (called Cariboo Castle) are still to be traced" (Grey 1858, plate of Cartwright community; also DeVolpi 1972, plate 38). We learn from various entries in the *Journal* that Caribou Castle had a storeroom, a small kitchen with windows, a dining room with windows, and an entrance porch; it undoubtedly had bedrooms and was of the longhouse type.

The Isthmus Bay houses — a temporary dwelling on Great Island and a more permanent one inside the bay — are also not described in the *Journal.* A brief note on the addition of a kitchen measuring sixteen by twelve feet to Robert Collingham's house in Paradise River shows a further adherence to the sixteen-foot measure (C.20.9.83).

In contrast, the building plans in the new Cartwright papers are for larger dwellings. For instance, the plan of a merchant house on page 33 of *Additions* is 132 feet, four inches, by forty-four feet, four inches (four inches may represent the thickness of the walls), while still keeping to the sixteen-foot dimensional measure. Its layout of rooms and use of space, the construction techniques, and suggested materials all closely follow the approaches described by Cartwright in the *Journal.*

The frame construction techniques described in the new papers, especially in the entry "A Neat Studded House" (*Additions*, 61), resemble those known to have been used in eighteenth-century England, early Virginia, and elsewhere along the eastern seaboard, and again correspond well with information found in the *Journal* (cf. also Mellin 2003, 98). House construction began with a softwood frame of sleepers and wall plate that held two layers of vertical studs placed close together to carry the external and internal boards. Insulation was by chinking, or "chincing," between the studs with moss or clay, and the whole was covered with heavy paper and sawn boards or squared logs. Roofing was secured by laying sheathing paper on roof boards with boiled tar and battening this layer with lathes or boards. A small number of references to roofs in *Additions* (19, 30, 33) suggest a ridge line rather than a flat or front-to-back sloping design, but not necessarily a central ridge, as suggested by one instruction that the "Ridge of the Roof ... be over the North Partition." A shallow slope is indicated in the instruction that "the fall from the ridge to the eaves" be two inches in one case and four inches in another.

Building foundations are infrequently described in the *Journal*. There is little evidence for earthfast-type construction, whereby supports were dug into the ground. Instead, the approach seems to have been that of a timber frame on levelled ground or on shores. In the entry "To build a House of Boards upon the Ground" (*Additions*, 63) Cartwright instructed that the ground should be levelled and strewn with a layer of stones or gravel and a layer of fine sand. The planks for the foundations of the outer walls and partitions would then be laid upon this prepared surface. This may have been the exact approach taken at Cartwright's fishing room at Stage Cove, near the mouth of the St Charles River, where archaeological work recorded low soil ridges that delimited a levelled surface closely matching the known dimensions of the house (McAleese 1991). The soil ridges were probably left in place to act as insulation around the bottom part of the house, an approach that Cartwright recorded for insulating the base of Caribou Castle (C.29.10.75).

Cartwright's construction somewhat resembles an eighteenth-century colonial built type known as *"piquet sur sole,"* referring to vertical supports on a sill plate, for which there are few detailed descriptions (Harris, ed. 1987, plate 56; S. Myers 2006, personal communication). A conservation architect's interpretation of this form based on Cartwright's "A Neat Studded House" is shown in figure 11.

Not all of Cartwright's dwellings were built with the foundation resting directly on the ground. The *Journal* tells us that "stouters, posts [and] shores" were used for the Isthmus Bay fishing room on Great Island (C.14.7.77; stouters being extra long and strong shores placed at the head of a stage), while his residence at Isthmus Bay had "some fresh shores ... put under the platform" (C.16.9.85). Posts or shore supports under a building allowed construction on virtually any type of surface with a minimum of ground levelling. They made it easy to move a building or to float it to another location, a common practice in the early fishing communities

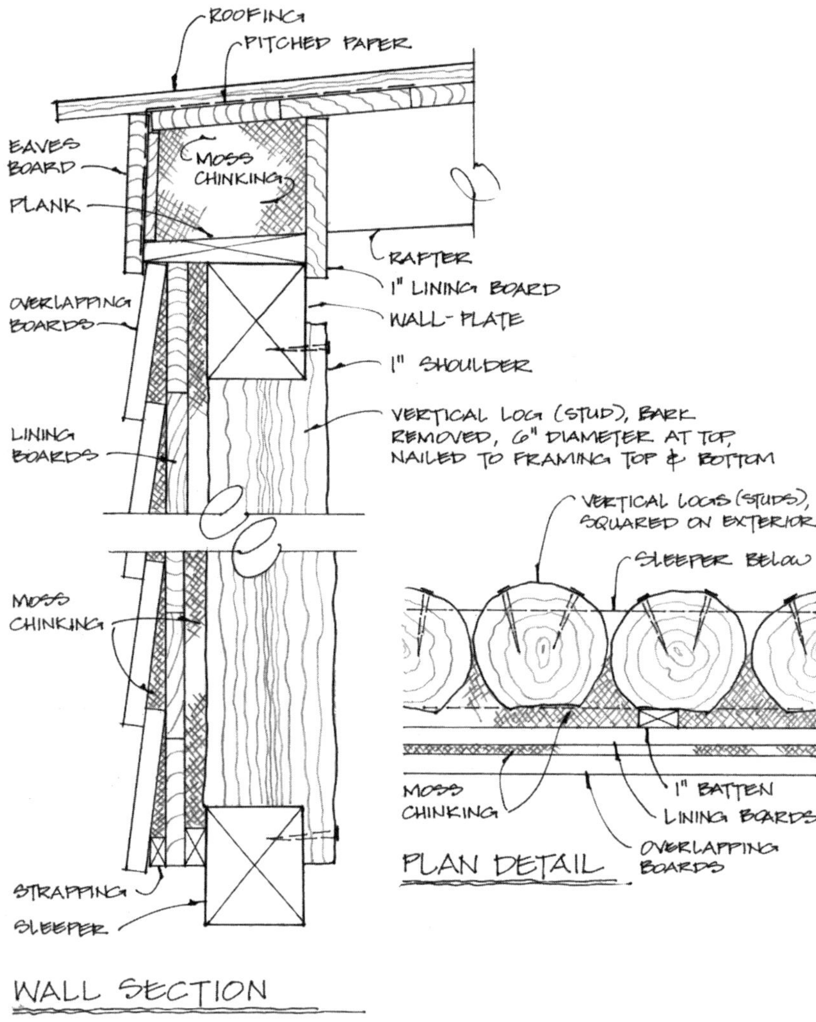

Fig. 11   An interpretation of Cartwright's "A Neat Studded House" from *Additions*, pg. 61, by Susann Myers, conservation architect, Public Works and Government Services Canada.

Fig. 12   Wood frame houses at Cape Charles Cove, Labrador, showing the shores used to support these buildings. Photo by M. Stopp 2003

of Newfoundland and Labrador where land might not change hands but houses could (cf. Mellin 2003). Cartwright referred to shores in the plan of a merchant's house in the loose papers. The use of shores to support a wood-frame house may also explain why archaeological testing at Ranger Lodge, Cartwright's first residence in Labrador, did not yield evidence of building foundations or a building surface. The natural terrain at this site consists of cobbles and boulders, which would have presented difficulties for laying a foundation but been useful as strengthening supports for shore posts (Stopp 2004) (figure 12).

One of the important contributions of the new papers is the detailed description of materials and the construction guidelines for early wood-frame buildings. Whatever their precise architectural origin, Cartwright's rectangular dwellings were functional structures designed to house an extended "family" consisting of the merchant and his domestic and professional staff, while the male work crews had separate quarters also of rectangular design.

CONTACTS WITH ABORIGINAL PEOPLES

Cartwright had high hopes of establishing lucrative trade relationships with both Innu (whom he referred to as Nescaupik or Mountaineer Indians) and Inuit (whom he referred to as Esquimaux Indians), but throughout his years on the coast his relations with the Innu remained intermittent. Among the Inuit, however, he succeeded in making unprecedented strides at a time when Inuit-European relations had been fractious for nearly a century. His links with Inuit were enduring, and he was still trading with them in his final years in Labrador. The *Journal* recounts many interactions with Inuit that included visits to their campsites, Inuit workers in his household, attempts to develop cross-cultural ties by bringing Inuit to England, possibly a son by an Inuit woman, and, just months before leaving Labrador for good, a spurned marriage proposal to sixteen-year-old Eketcheak, who considered him far too old and unsavoury at forty-eight.

The merchant firm of Noble and Pinson unsuccessfully tried to wrest the trading branch of his business from him when they acquired the Sandwich Bay properties, but Cartwright refused to relinquish this small part of his enterprise. By this time in the 1780s there were many merchants, including the Moravians, competing for the Aboriginal trade. In one of

Cartwright's last *Journal* entries we learn that his partner Robert Collingham was still involved in trade and made the journey from Isthmus Bay to Groswater Bay in order to obtain "a small quantity of oil, whalebone, and skins" from Inuit, and that competing interests from Quebec and England were already set up there (C.19.6.86).

Cartwright's entry in *Additions* entitled "Eskimeau Indians" (93) is obviously based on his personal experiences. It sheds some light on the nature and considerations of a trade enterprise while also reflecting a generosity of spirit towards indigenous peoples that seems to have developed over time. In an early letter to his brother John (20 September 1771; see page 215 below), he criticized the Inuit as "a sorry race of mortals" and was mortified by their social arrangements, their way of eating, and their sanitary habits – in short, by everything about them. His later writings, however, show a developed respect for their skills and knowledge and an adjustment of earlier impressions. His instructions to "observe strict honesty in your dealings: sell them no bad or damaged goods" was undoubtedly a rare dictum among merchants of that time. His further advice, that "as soon as you can discover the leading men, attach them to your interest by associating freely with them and admitting them to your table at times ... and you may depend upon succeeding to the utmost of your wishes," admits to rather more commercial cunning.

Two lists of goods "for the Eskimeau trade" describe the many items sought by Aboriginal traders in exchange for their baleen, whalebone, furs, and oil. One of these lists appears on page 96 of *Additions*. The other list, presented here as figure 13, was not part of the Cartwright family archive but formed part of a 1783 letter to Robert Hunter who was Cartwright's factor (or assignee, or agent) in London; it was found by I. Marshall at the Dorset County Records office. Both lists share similar items of hardware, and the 1783 letter also shows quantities. A 1771 letter to his

Fig. 13 (*opposite*)   A list of items for trade with the Inuit. This list is not part of the new papers but is excerpted from a letter of 3 February 1783 written by George Cartwright at Marnham to his agent Robert Hunter. DRO, Lester and Garland Papers, D365-letters; some items are preceded by check-marks not duplicated here.

In order that you may make preparation for the Eskimeau trade I here under send you a list of such goods as will be wanted.

5 Doz. Arrow heads

5 Doz. Dart heads

3 Doz. Ullows

{I will get patterns made here and give them to you when we meet in town.}

5 Doz. Knives — the size & shape of carving knives made to rivet, but no handles

5 Doz. Ivory combs (the short ones about 3 inches long)

6 Doz. Of Taylors thimbles

1000 needles (fine long darning kneedls [sic])

20 lb of Bugles (round glass beads rather less than double Brisket shot the coullours [sic] Black, Blue, Red, & White)

4 pieces of Swanskin

2 Doz. boats kettles (the larger the better)

3 Doz. small Hatchets

1 Doz. Handsaws

9 Doz. d°. files

12 small Furriers guns about 16/ each

1 [hundredweight?] Powder

4 [hundredweight?] Shot

500 Flints

5 Doz. [?] Stockings, such as mine the best sort and I will order them on receiving your directions so to do.

20 Blankets. Cannot be too thick, or too course, & the larger the better.

4 Skiff roder [?]

2 Quoils of two inch rope {if twice Paid}

2 d°. of 1 ½ inch d°. {if twice Paid}

1 B^l. of Pitch

1 d°. of Tar

three pairs of caulking irons for shallops

3 caulking mallets

3 Doz. Beaver traps

A few old Whaling harpoons & spears if they could be had but particularly the spears (what they kill the whales with after they are stuck).

father, William Cartwright, given in the correspondence, explains that beads were a staple trade commodity. Their popularity is corroborated by archaeological work at Ranger Lodge where an exceptionally large number (1,896) of black, blue, red, and white glass beads were recovered (Stopp 2004).

Beads in these same colours come to life in the watercolour portrait of an Inuit woman found among the Cartwright papers in a scrapbook belonging to Frances Dorothy Cartwright, George's niece and the author of a biography of her other uncle, John, entitled *The Life and Correspondence of Major Cartwright* (London, 1826). The portrait (figure 14) is of a young Inuit woman in traditional garb decorated with red, white, blue, and black beads. The artist and subject are unknown, but a few deductions can nevertheless be made about this picture. The woman's head band, beaded ear pieces, and the looping beadwork pattern on the amauti she is wearing, and the amauti's long, ovulate tail, suggest that this may be Caubvick, who travelled with Cartwright and a group of Inuit to England in 1772. A full-figure portrait of her was drawn by Nathanial Dance at that time and shows similar design details but is not in colour (see note 7).

The artistry of this picture is amateurish yet precise and may have been executed by Cartwright himself. It could also have been painted by one of his family; his sister, Catherine, is known to have befriended Caubvick during the visit, while Frances Dorothy, into whose scrapbook it was pasted, may also have been the artist. The careful rendering of alternating bars of red, black, and white beads on the front of the amauti and the ear pieces, and the checkered pattern in blue, black, and white beads that edged it and the tops of the leggings are details that suggest that the artist had seen this woman. Her leggings in red may be of European fabric or may be an attempt to convey the deep reddish-brown of tanned hide. A rare portrayal, the drawing shows a blend of skin clothing of traditional cut decorated with much European beadwork that would have been characteristic of Labrador Inuit who had access to European trade goods – a combination that would largely disappear in the 1800s as women switched to European fabrics and designs.

Cartwright may have given poorer trade value for Aboriginal goods than other merchants. His suggested prices for certain items are found on page 106 of *Additions*. Prices from the island of Newfoundland in 1763,

Fig. 14   An Inuit woman wearing a beaded amauti and beaded ear pieces. Original is a 10 cm by 8 cm watercolour pasted into a scrapbook of family mementos and news clippings belonging to Frances Dorothy Cartwright. Reproduced with permission of John Cartwright; in private collection

in comparison, were much higher than those offered by Cartwright, who suggested five shillings for a prime silver fox and one shilling for an otter skin, as opposed to a guinea for the former and ten shillings for the latter paid on the island (Marshall 1996, 487n30).

In developing trade relations with Innu and Inuit, Cartwright was exposed to the ways of wholly new cultures. In *Additions* he recorded a number of observations of ethnohistoric interest from both groups. These include a description of the Innu way of rendering animal fat; an important nutritional requirement for all northern peoples, fat in rendered, solidified form was easy to transport and preserve (99). On the same page, the Innu technique for skinning an otter is described as being performed by women using "the long bone in the fore-leg of a White bear." Very appealing is Cartwright's brief description of an Innu tobogganing pastime, embedded in the note on the sailing sled (104). A 1771 letter to Anthony Eyre (after whom Cartwright named Eyre Island, today's Assizes Island at the mouth of the Charles River) contains important early descriptions of an Inuit snow house, soapstone lamps, and sled dogs (descriptions of the same are found in the *Journal*, C.26.2.71).[17]

Cartwright adopted a number of Aboriginal strategies while in Labrador, among them the use of Inuit snow goggles or "snow-eyes" (C.19.2.72; 19.2.86). Early on he adapted to the kayak – "The Indian [Inuit] women began to cover my kyack with new skins" (C.10.11.75) – and appreciated the heat-retaining benefits of a caribou-skin sleeping bag over woollen blankets while camping (C.12.2.78). For winter travel he adopted Innu snowshoes and toboggans and Inuit sleds (C.18.2.72, C.24.2.72, C.8.1.75, C.1.4.86). In *Additions*, his instructions for sleds (81, 101, 103) are based on the use of "Nescaupick and Esquimau" sleds recorded in the *Journal* (C.5.5.72, C.1.3.75, C.11.4.75). The entry in *Additions* (52) entitled "An Improved Way of Jerking Salmon, Codfish, Etc." characteristically combines Cartwright's own innovation (in this case, a substantial drying frame) with a technique learned from the Inuit for making *pipshy*, or dried fish, that he had learned early in his Labrador years (C.2.7.71). His portrait (frontispiece; figure 1) by W. Hilton is a good example of the amalgam of European and indigenous adaptations. Cartwright wears a traditional Innu man's painted caribou hunting coat and sash, Innu snowshoes, and Inuit sealskin leggings and footwear alongside a European hat (of beaver), a Hanoverian rifle, and a hunting hound.

CONCLUSION

Few people today will be familiar with the technologies described in the new Cartwright papers from the vantage of personal experience. Those who still remember the intricacies of placing a sealing pound or the proper construction of a deadfall trap are today's oldest generation of Labrador residents. Cartwright has left information that will contribute to an understanding and appreciation of the historic depth of a nearly forgotten resource-based way of life and the breadth of traditional technical and ecological knowledge possessed by the people of Labrador.

Cartwright wrote many of these new papers at a time when enterprise in Labrador and in eastern North America in general was on the increase. He seems to have identified a niche for his knowledge that included merchants and workers who had their sights set on Britain's northern colony, as implied by "Memorandum for W" and by two letters in the correspondence from the HBC and Dr Grenfell, both of whom were interested in Cartwright's instructions.

The new Cartwright papers give an indication of the material culture complexity of coastal frontier life and represent the acquired knowledge of someone whose hands had built houses and traps and who had spent winter days and nights on the barrens. It was through his will to learn and his diligence that Cartwright was able to savour everyday satisfactions such as "the window was fixed in the store-room, the sod wall was finished [around the house], and I had a hundred and five pieces of venison hung up in the kitchen to smoke" (C.2.10.75). There is no doubt that he made many contributions to the early resource industries along the coast. His plans in the new papers for sealing pounds, house building, and trapping were based on methods used in Labrador, while the technical details and material lists anchor all of these instructions to experience. Given his will to learn, to adapt, and to improve, it comes as no surprise that he also wished to record some of this hard-earned knowledge, which was perhaps his most enduring outcome, his only profit, from his years on the coast of Labrador.

PART TWO

The New Cartwright Papers

Cartwright's table of contents for Additions to the Labrador Companion *was arranged as two columns. The page numbers given are included in the transcribed text. (See table 3, page 37, for the table of contents rearranged by subject.) The reader is reminded that all transcriptions are based on microfilmed copies of the originals, which is especially apparent in the sketches where number keys are too faint to read (or missing) and have been added.*

[pg. 1]

To catch any Beast which is in an Earth, or their young ones when old enough to come out

Place a Box trap, of a proper size, close before the mouth of the Earth, and build up to it with ſtones, or such other materials as the place affords. Spirit of Vitriol, poured upon it all in the mouth of windward hole, will force them to Bolt.

To Catch Otters upon a Path

Where two Bays, Harbours, Ponds or other waters lay over to each other and do not communicate by a ſtream, the otters are sure to have a path across the land from one to the other, and as they associate in great numbers so soon as their young are grown ſtrong enough to follow their Dams and fish for themselves, you may catch the whole of those who shall attempt to cross from one water to the other at one time by the following contrivance.

Where the ground is moſt suitable for the purpose, and as near the centre of the Path as may be, fence the path on both sides at two yards distance from it for a length not less than sixty yards. Fence the ends in also, within one foot on each side of the path. Make a door to hang over the ends of the path of 4 feet in height (and which muſt be the height of the whole of the fence) with a Beam across from one side-fence to the other, three inches before the bottom of the Door when raised up to an horizontal position. Fix a hooked ſtick upon that Beam to keep the Door elevated; fix a wire to the back of that hook, and lead it along through ſtaples or rings fixed along the side-fence to the centre, where the other end muſt be made faſt to a bridge, and fence the centre across. Let the otters enter at which end they will. When the firſt comes to the centre he will tread upon the bridge, draw the hook from under the Door and it will then fall down; by which the whole flock will be secured for they cannot get over a fence of four feet in height. A line will not do so well as wire because it will slacken in dry-weather and contraƈt in wet and consequently either let the Door fall by its contraƈtion, or not do so when it is enlarged. Each entrance should be blinded by ſticking green boughs close to the outside of the end fences. [pg. 2]

To fit up a bladder, for the purpose of giving bouyancy to a Boat, Raft, or any other thing

Cut out a bladder with as long a neck as possible, and take off what fat
or flesh may be upon the neck; blow it up, and tie it near the extremity.
Provide a Spiggot and Faucet. Saw off about an inch and a half of that part
of the Faucet which shall best fit the cavity of the neck of the Bladder,
and saw several rings round it, sufficiently wide to admit the twine you
intend to lash it on with; fit the Spiggott into it, saw the projection of the
small end off close to the end of the Faucet and the large end half an inch
beyond the largest end of the Faucett, and flatten the sides of it. So soon
as the neck of the bladder is half dried, smear the outside of the Faucett
with white paint, introduce it into the inside of the bladder neck and
would [wind?] it close round from end to end; fixing a sufficient length
of twine on each side near the lower end to tie over the end of the Spigott,
and lay the lashing well over with black sealing-wax dissolved in Spirits
of wine. When that is dryed, which it will be in a quarter of an hour, blow
the Bladder full of wind and take a few turns of twine close under the end
of the Faucet until you have fixed in the Spigot and tied it in, when you are
to take off that twine, that the neck may expand to its full size. At future
times; when the Bladder will be gotten perfectly dry, lay the neck in water
until it is quite soft and pliant before you blow it up. In case you want to
keep the Bladder constantly blown up and never let the wind out, smeer
[*sic*] the Spigott with white-paint before you fasten it in, and then it
can neither come out again nor can it lose the wind. A painted cork
introduced in the neck of the bladder, when blown up, and lashed
in will do.

To construct an excellent portable Back-tilt; two of which placed facing each other will make a capital whigwham

Provide a sufficient quantity of oiled silk, or seal's gut, as well make a
sheet twelve feet 6 inches long by ten wide. Sew a tape across one end six
inches short of it, and another tape across the lower part, eight feet six
inches distant from the former. The Rooftree, Rafters and Walplate must
be made of Saint Peter's line in [pg. 3] the following manner. Sew one
piece across the top upon the tape fixed there, and splice an eye at each

end of it; sew another piece upon the lower tape and splice an eye at each end of that also: you are then to fix other pieces of lines from the top line, (which is to be the Rooftree), to the bottom one, (which is to serve for a Walplate), at one foot distance, upon tapes previously sewed on upon the covering (one of which must be upon each edge of it), and those are to be for the rafters. By additional lines suspend the top-line (or Rooftree) to two trees at the height of seven feet, and suspend the bottom line in the same manner at the height of three feet. You are then to fix other spare lines to eyes spliced at the bottom of each rafter-line and peg them into the ground at proper distances behind them in a direct line with the two cross lines. There will then be six inches of the top of the tilt hang[ing] perpendicular down, and also three feet three inches of the bottom, and you must then sew on other lengths of the same materials to form the ends, as represented in the plan underneath.

Elevation of the end of the Tilt

N.B. If two eyes were fixed upon the Ridge-line at 3 f. 4 inches from each corner of the front of the Tilt, it might be guyed out in such a manner as to prevent its baggin[g] in the centre. Seal's gut is excellent for this purpose, and may be procured in any quantity by those who have a Sealing-post. It must be sewed with narrow straps of the same, but with a [base] of [support] Line.

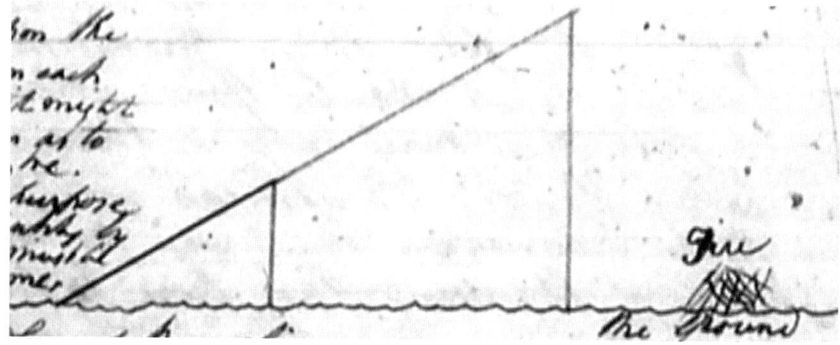

Plan of the Lines

N.B. Any two trees will serve to extend the Ridge-line but it will not be
likely that [an]other two will be found to ſtand at proper diſtances for the
Walplate-lines, but that may be easily contrived by fixing Guys to the eyes
at each end and guying it either backward or forward until it is brought
parallel with the Ridge-line. If a log of wood is laid upon the heads of the
Pegs it will keep them from drawing out of the ground. N.B. The lower
tape muſt be broad, with Eyelet-holes in it for the ends of the Tail lines to
go through.

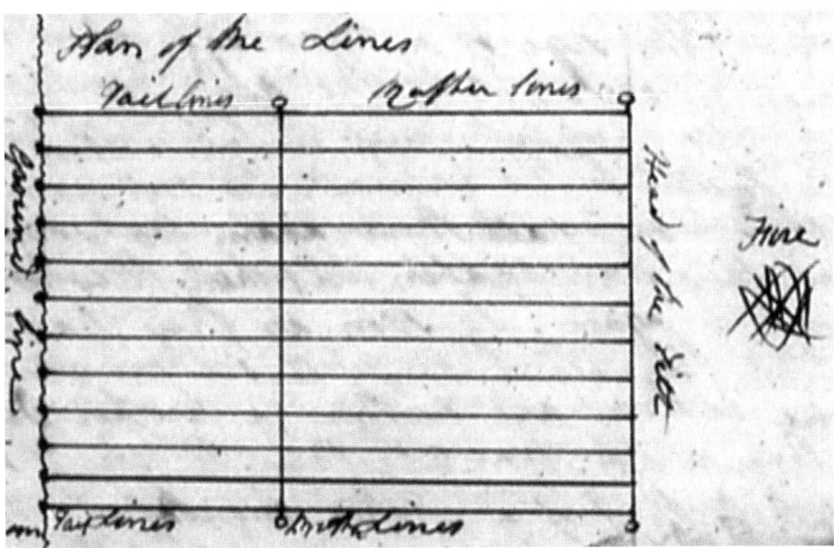

[pg. 4]
**To fence a Salt-water shore in such manner as to direct animals into a
Pitfal-trap built upon the Beach**

Set a line of Beams upon Poſts and Shores; the outer end of which muſt
ſtand in the water below the fall of the Tide, and the inner end be fixed

close to the Trap, and rear trunks of small trees against it. All those which stand in the water, when the tide is in, should be nailed to the Beam, or otherwise fastened, and their but-ends secured with large stones. The outer ends should project to a proper distance in front of the trap each way.

To teach long-winged Hawks to wait on well & close

As soon as they are old enough to fly, bring them from the fist of one man, to the Lure thrown out by another, at the distance of about twenty yards at the first, and increase the distance daily, as they get strength. For a few of the first times throw the Lure out on their first coming to it; baulk them afterwards for one or two times, that they may take a turn or two before they are taken down. As soon as you perceive them strong enough to keep upon the wing for any length of time, instead of throwing the lure to them, toss up a bit of meat the size of a hen's egg, or a small bird or Partridge head, when they are near enough to catch it in the air. For the first few times they will probably miss it, but will pitch upon it after it is fallen to the ground. Do not go up to them, but walk on and call them again so soon as you observe that they have eaten it, when they will rise and follow you again. So soon as they have caught one bait in the air, they will seldom miss doing it every time, if they are within a proper distance when it is thrown up. You are then to reduce the size of the meat to that of a proper mouthful; keep walking on, and toss up other bits, from time to time, until you have [pg. 5] given them nearly their whole meal, when you must take them down with the Lure, or they will go to tree, and be a long time before they will come down. As they gain more strength, keep them longer upon the wing between each bit of meat. So long as they wait well and close on, which they certainly will do by this method of treatment, do not call to them; but do it as often as they rake off, and toss them a bit of meat so soon as they come in.

By pursuing the above method you will observe, that they will not only wait on close, but also fly low. Although that is a fault, do not offer to break them of it until near the time that you intend to enter them to Game: you must then do it by tossing up live Larks. Pull a couple of feathers out of one wing of the first two or three Larks, that they may be sure to catch them, but do not do that afterwards; for when they find the disadvantage

of being low, when the Lark is thrown up, they will instinctively take a
lofty station. If young Hawks are too soon taught to take a lofty station,
they are apt to fly too wide, and then are seldom well laid in on the rising
of the game. For two or three days before the game season commences,
procure a few live ones, and toss up one to them each day; but if you
cannot get any, a chicken of the same size and color, must be substituted.
The disadvantage of chickens is, that they cannot fly; and therefore
always fall to the ground before the Hawk can catch them: and a long-
winged Hawk is afraid of striking at a bird upon the ground, lest she kill
herself against it: but short-winged Hawks kill as well upon the ground as
in the air.

The Lark must be [sieled], which is done by pulling a feather out of its
tail, stripping the plume off one side, then put it through the underlid of
one eye, over the beak & then through the other under eyelid, with the
plume side down-wards. [pg. 6]

The best way to Mew an Eyass[4] Hawk of either Kind

When you leave off flying in the Spring or the Summer, clip one wing of
your Hawk and turn her into a Garden or other place securely fenced all
round, lest she get out and be worried by Pigs. Go to her frequently, the
oftener the better, sit down upon the grass or a low seat set upon it, and
feed her upon your fist with hot meat as often as you can get it, and the
Pigeon-cote will supply you. Stroke and handle her very often, and slip
her Hood on and off several times every Day. By not being able to fly she
will become tamer than usual, and as she will cast only one feather in
each wing at a time, and be nearly a month between one pair and the next,
she will not be able to fly until the end of the Summer, and then suffer her
to fly at her own discretion until she is full rimmed; by which time she
will have enseamed herself and be fit to fly at her Quarry. If there is not a
proper place for her to bathe in, you must sink a proper bathing-tub in the
ground. So soon as she gets the use of her wings, keep her waiting on, as
directed in the foregoing page, and you may be certain that she will do it
well and not desert you, unless you neglect to feed her as often as is necessary. Blocks and Perches should be placed under a shed or a tree for her
to sit upon and be sheltered from the weather. When a Hawk is mewed in
a Room or upon a Block, the want of proper exercise injures their health,

which is the cause of their not being fit for use more than three or four years, but they will certainly laſt longer if treated as above direćted; and will be in proper condition to [pg. 7] fly a month or six weeks sooner, as well be as strong as any haggard.[5] They will also be in no danger of being over-heated in their grease, or not being properly enseamed: for they will manage themselves much better than you can do, by taking what quantity of exercise and number of stones are necessary, and no more. A Haggard might be treated in the same manner until she gets a sufficient number of new feathers to enable her to fly but must then be confined, or she would soon provide her own food and consequently leave you. I am not certain whether an Eyass which had been flown at Crows the season before might not do the same, but there would be no fear of one which had been flown at Game only. Never give an Eyass a Pigeon in the feathers, for that is a sure way to dispose her to fly at wild ones, and render her unfit for your use. If there is a pond in the garden, which is too deep for her round the edges, it should be fenced off by pricking up sticks round all such deep places; as she will be drowned, if she goes in at them.

N.B. If your Hawk absents herself for a Day or will not come down to the Lure, after she has got the use of her wings, you may be certain that she has procured her own food, and if you do not then confine her, she will more probably leave you in a short time after.

To fix Salmon racks across a River, in such a manner as effectively to prevent any large fish or otter getting under them

Throw a Bridge across the River, consiſting of two sets of Beams supported with Poſts and shores as direćted on Page 326 of the Labrador Companion. If you cannot find long ſtones fit for the purpose, make Killicks with claws no longer than sufficient to fix the rods in them, and fix one of them close to the foot of each Poſt and on the lower side of the ſtream, to prevent the heel of the Poſt from being displaced by the pressure of the Water. Each [pg. 8] Post must be at the exact length of one Rack from each other that the heads of the Racks may rest against them. Make a sufficient number of Racks to reach across; and the lower ends of each head of them must be six or eight inches lower than the under bar, and pointed, if the bed of the River is sandy or muddy, that they may be driven into it; but flat if the bed is Rocky, and the heels of the Post must be cut in the same manner. The two bars of the Racks having been bored full of holes at proper distances from each other, the staves are to be put into them after the frames are fixed in their places, and the lower end of each thrust down into, or upon the bed of the River; which will stop every inequality of depth that there may be and prevent fish or otters from passing under the lower bar. Both Bridge and Racks must stand at such a height as to be above flood mark, or they will be carried away in case the water flows over them. If the bed of the River is soft enough for the heels of the Bridge posts to be driven into it, there will be no occasion to place long stones or Killicks behind them.

To construct a Bridge across a River or other Water for foot passengers to cross upon

Provide a sufficient number of ½ inch Planks; make a hole in each end of them large enough for an inch rope to go through; nail two pieces of sheet Cork under each end. Provide an equal number of Killicks with an inch rope to each for a mooring. Moor as many of those Planks within a proper distance of each other as will extend across the river and support the end of the mooring with a lag-buoy, as practised with Salmon nets, and then nail two rows of 1½ inch Plank across the River, their ends resting upon those Planks which are moored up & down the stream. They must be taken up before the River freezes over or they will be carried away by the ice when that breaks up. If such a bridge is constructed across a dead water, a lower mooring will be necessary, or a gale of wind will break it: and that will also be the case on a River, unless there is a strong stream. Shaved logs of dry [wood] will be better than the lower set of planks. [pg. 9]

To construct a Bridge across a River, for the Passage of an Army, in lieu of one of Pontoons

Provide a sufficient number of Casks (say Tierces). Make no bung-hole, but there must be a small hole in one head of each for [blowing] them off. Hoop them with iron, and the inner hoop upon each chime must be very strong, with a staple well riveted in one, and a hook in the other, and they must stand opposite to each other. A chain must be fixed in each staple. Place two of these casks parallel to each other, with the hook of the one opposed to the ring of the other, and fasten them together at a proper distance asunder by the spare end of the chains being hooked upon the opposite Cask. Having fastened two other Casks in the same manner, lay the ends of a Beam upon the Chains of the two pairs of Casks, and tie it fast to the rings and staples of them. Bore an augur-hole through each end of these Beams, for moorings to be served through, and moor a sufficient number of them across the River, at both ends, in the same manner as Pontoons are moored; then lay other Beams across, from one to the other, and nail Planks across them. The size of the Casks must be determined by the number of them, and the weight they are intended to support, and the less they are, the less chance there will be of their being struck by a shot from the Enemy. As the Casks will be apt to dry & leak, by being kept long out of the water and in the sun, they must always be [blown] off before they are used, and a Cooper in readiness to harden on the hoops when they require it. In case any of them should be struck by shot, they must be replaced by others, which will not be more difficult [pg. 10] to do, than the replacing of a Pontoon. By this invention, a very considerable saving of expense will be made, for the Casks will not cost so much as Pontoons, nor will they require so many Carriages, Horses, or Drivers; much less Forage will be required, and they will be repaired at a much less expense. The Beams which are laid upon the Casks might be secured in their places by passing the chains through staples driven into the lower sides of them, but then their ends could not so easily be got out, in case they were broken by a shot, and therefore I think the other way preferable. It might be best to fix two pairs of Casks at each end: for if one of them only was damaged by a shot, the other pair would support the Beam, and thereby make it much easier to replace the damaged one. The Casks will not be more liable to be struck by shot than Pontoons are.

And if a short block of light wood is nailed upon the upper side of each end of the cross Beam, and so thick that the upper side of it shall be one inch higher than the upper side of the Casks, the long Beams may be laid direct over the centre of those Casks, and as the ends of the Planks will project over the outer chimes, they will be compleatly under the Bridge, and less liable to be struck by a shot. The length of the chains will be a means of replacing a damaged cask with more facility, in case another set of Beams should be required to be laid across the centre of the cross Beams. Logs of wood must be fixed [pg. 11] there likewise. By this method, other pairs of Casks may be fixed under the Bridge, to give what Buoyancy may be required. The more Casks there are under each cross Beam, the less inconvenience would arise from one being destroyed: and it may be thought better to have each cask attached to the Beam independent of its fellow, by having two chains fixed upon the Beam with a hook at the end of each, by a staple under the centre; and staples only upon the strong hoops. It will be no difficult matter to replace one that is destroyed, and it will be much more difficult to break such a bridge, than one constructed upon Pontoons.

Upon reconsidering the above I am of opinion that instead of Chains under the Beams, a bar of iron had best be nailed under them with a hook at each end projecting its own length only beyond the sides of the Beam and turned downwards, to hook into the rings upon the rope hoops of the Casks: and staples should be driven into the upper side of the Beam; direct over those bars, with the top of it even with the surface of the Beam & the wood cut away, to admit of a hook being fixed in it, that a lever may be fixed to press down a fresh Cask which may be wanted to replace one that may be damaged.

To catch a Seal with hooks

Make a strong plat of horse-sinew, of a foot long, or more; whip one end of it for six or seven inches, with fine waxed twine, and fix a strong swivel at the other end. Tie three of the strongest Irish Pike-hooks upon the whipping, one above the other and a Whiting-hook opposite to each. Splice two Bank lines together, and splice the end of one of them into the eye of the swivel. Reeve the other end of the line through a ring fixed upon the stem of a Skiff, and tie it to the aftermost Thwart. Bait with a small fish, such as a Herring, Trout or any other, by fixing the Whiting-hooks into [pg. 12] it, as if for Pike-fishing, with the tail bent to make it spin. Hale in as much of the line as will bring the bait within a proper distance of the Boat, and row away at the rate of not more than three Knots. When a seal seizes the bait and is hooked, let him tow the boat until he is tired, and you may occasionally hold against him with the oars. As he tires you may hale in the line, and vere again as necessary, and a wooden pin fixed in the thwart would be very convenient to save your hands in vereing and would produce greater resistance. By playing him in that manner, as an expert angler does a salmon, you will tire him compleatly at last, and kill him with a bludgeon. As seals are much disposed to follow a Boat, I have no doubt of their taking the bait, nor of their being killed in that manner, if properly hooked; a Bank line being strong enough to tow a light skiff.

Posts or Studs which are to be fixed in the ground

To preserve the bottom end, which is within the ground from rotting, dip it, for a few minutes, in a Pot of boiling Pitch.

General directions for a Sealing-Post

As Seals always migrate from the north towards the South, that they may find an open Sea during the Winter, the proper places to catch them at are either on the South side of a Bay and near to the mouth of it, or between an island and the Continent upon a straight shore. It must be sufficiently exposed to the sea to prevent its being frozen up soon, but not so much exposed to it as to cause the sea to break so heavily upon the shore as to wash and tear the nets; and if it is formed by an Island, that should not be at a greater distance from the Continent than sixty, or at the most seventy fathoms, nor less than twenty; nor should the depth of water be more than six fathoms, nor less than three.

[pg. 13] Nevertheless a Post may be established where an island lies at a greater distance, but then it must be done in the same manner as where there is none, by the assistance of a Barrier net, moored between the Continent and the Island; parallel to both. A short Sealing-post cannot prove a good one, because there is not extent sufficient to fix a sufficient number of nets in it, but if it is too long, it may be apt to freeze up too soon. One hundred yards in length will do tollerably [*sic*] well, but it had better be one hundred and fifty, and I consider four fathoms of water as the best depth; as also thirty fathoms the best width. The Spring-tides upon the Coast of Labrador rise about six feet and the neeps about four feet and a half. Whatever is the length of your Sealing-post, have two tail-nets at the extremity of it to the Eastward, or Southward, according to which it runs; (observe that I am speaking of that part of the Coast which lays between the Straights of Belle Isle and Hudson's Bay) place them one foot more asunder than the length of your skiffs, heave them tight up and buoy them sufficiently to prevent their being sunk by the weight of Seals, which may strike into them. Next to the Tail nets must be the Murderers, which must be placed at the same distance from each other; but they must not be buoyed unless it be part of the farther end of each when they exceed thirty fathoms in length, as they must be lowered and hove up again as occasion requires to let the Seals in between them. Next to the Murderers must be the Stoppers, and the distances between them must be about six or seven fathoms; and the Entry net must be placed at the western or northern extremity of your Post, and should be about twenty fathoms distance from the first Stopper. All those nets are to be lowered occasionally, and therefore every net must have a Capstern [capstan] fixed opposite to the inner end of it, to do that by, and be sure to make every net full six feet deeper than what the depth of the highest tide will require. The greatest errors which I have observed in catching Seals, and which [pg. 14] were committed at every Post that I have yet seen, were the grudging the expense of a sufficient quantity of Twine and Cordage, both in the number and depth of the Nets. By not allowing a sufficient number of nets, the Pounds are too large, which gives the seals so much room, that they will not strike into the nets without firing a great deal of Powder at them; that is expensive, takes up time, and turns many shoals back. By making the nets no deeper than the depth of the water at the height of the Spring tides, so soon as a few Seals are meshed, the foot-ropes are tucked up and the greatest part of the rest pass under. Be assured that you cannot have too many nets, provided there is the length of a skiff between them and that an additional fathom of depth will repay the addition of expense with great interest. A weak Crew is also a very bad saving. You ought to have at least one boat for each net, two men for each boat, two or three more to attend the Capsterns and a Cook. There should also be a few spare nets, and one or two spare skiffs lest an accident happens to any of those in use.

The sixteenth Day of November is the Day on which the first Seal is expected and the whole are generally passed by the twenty second of December. The earlier the shoals of Seals appear, the greater success you may expect, and the reverse when they come late. When the Seals are arrived in greater plenty, if the wind chops about into the Southern quarters, they will not hurry forward, and the season will consequently prove a long and a profitable one, but the reverse must be expected if the wind holds to the Northward with severe frost. The number of nets will depend upon the length of your rope. After fixing the entry-net and three

Stoppers, the rest must be filled up with Murderers and the two Tail-nets, and the more you have, the better they will pay you. [pg. 15]

Stoves for warming a Kitchen & also for Cooking upon

	F.	In
The Length	2	8
Width	1	8
Height	1	4
Diameter of the Funnel	"	6

With a door for the funnel and a long blowing-hole. Each must be furnished with two Brass pans, or Iron Kettles, and they must be 18 inches long, 12 inches wide, and 10 inches deep, with covers. There must also be smaller ones, sauce-pans etc.

For Roasting

An iron Cabouse will be preferable to a Brick-fireplace; it must have a Funnel 8 inches in diameter; a false back, two short potcranes, and a leaf to hang down from the top of the front to prevent its smoking. The dimensions to be as follows, viz.:

	F.	In	
			N.B. There must be a spare
			false back
Height	3	[?]	
Width	3	[?]	
Depth	2	[?]	
Breadth of the leaf	[?]	3	fixed with hooks and rings
another leaf	1	8	N.B. The two leaves are to be shifted occasionally

A Bolting-yard to a large Pitfal Trap for winter use

Build it of Posts and Rails. The Posts must be only six feet asunder. The Rails very strong and only eight inches above each other, and the uppermost nine feet high. Have a net six feet deep and the mesh 4 ½ inches square, to place at the back of the Rails, to catch Foxes in if they should bolt also. Such a yard will not fill with drift as a studded one would do. A circular Trap would be best on any place exposed to drift, & the Bridges to it open underneath. A Trap seven feet high will require Bridges of ten feet six inches in length. [pg. 16]

Plan of a Sealing-post in the Harbour formed by Stoney and Venison Islands on the East Coast of Labrador

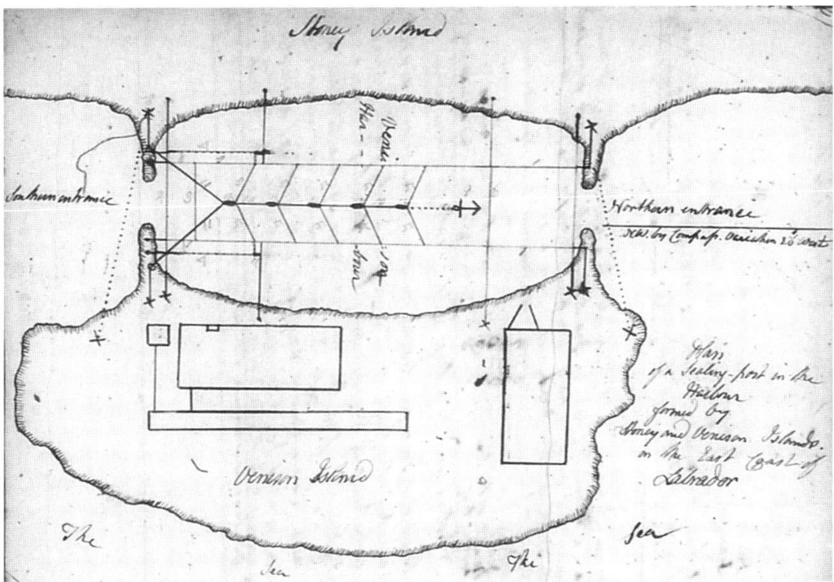

[pg. 17]
Explanation of the Plan on the other side[6]

1. The two Barrier nets, which are supposed to be two hundred yards long. Their head ropes are made fast upon the points of land at the South end of the Harbour, round through Blocks fixed upon the points at the North end, and hove tought [taut] by Capstans, expressed by crosses.

2. The outer Tail-net. The head-rope made fast upon Stoney Island, and hove tought by a Capstan fixed upon Venison Island. The length about 50 yards.

3. The inner Tail-net. Stopped upon the head-rope of the two Barriers, and two yards at each end stopped along them to the Northward, with a

stone fixed to each lower Clew[7] and the ends of it stopped in two places down the Barriers, to prevent the Seals from passing it. Length 40 yards.

4. The two Savealls to catch such Seals as may break through the Barriers, as there will be a great pressure at that part. The mooring for them are to be made fast to the head-ropes of the Barrier and to the shores. Length of each of these nets 44 yards.

5. The Hawk-nets. Their North ends secured to the two Barriers, and their south ends made fast to the Buoys upon their mooring. Their lower Clews to be tied together and a stone fixed thereto; the angle of the Southernmost pair is to be left open from top to bottom, and also that of the Northernmost pair, but only six feet at the upper part of all the rest. Length of each Hawk 20 yards, and they are to be fixed about 20 yards from each other.

6. The Stopper-net. The outer end of its Head-rope is made fast upon Stoney Island, and the inner end passed through a Rachet-bore fixed in an upper Mesh of the East Barrier and then hove tought by a Capstan fixed upon Venison Island. The length of it will be 36 yards. The outer end must be secured on to the west Barrier, and a stone fastened to the inner, bottom Clew.

7. The Entry-net. Its head-rope fixed in the same manner as the former, but this net must extend from shore to shore. It as well as the outer tail net will stand about five yards within the Harbour, that they may not be washed to pieces by the run upon the shores in Gales of wind. Its length will be 56 yards.

All the nets must be made six feet deeper than the depth of the water at the height of the Spring tides.

The Capstans, to heave the mooring tought by:

> The Anchor of the Hawk-mooring
>
> The Hawk-mooring
>
> The Guys of the Hawk-mooring, to keep the south end in the centre of the Harbour.
>
> The Buoys upon the mooring. Made of dry Fir; 4 feet long & 8 inches diameter.

N.B. Such parts of the Head-ropes and mooring as have no nets upon them are represented with black lines, and with red where there are nets.

Dwelling house. 73 feet long, 25 feet wide & 9 feet high with a break-weather [?] feet.

Oil and Skimming-house 75 feet long, 30 feet wide, 10 feet high, with a Chamber 7 feet high.

Rendering vat 120 feet long, 8 feet wide & 8 feet high.

Washing vat 10 feet square & 6 feet high

A spout to convey the Oil from the Rendering into the Washing vat.

A spout to convey the ran-fat into the Rendering vat. [pg. 18]

Venison Harbour

Arthur Thomas Slade, of Poole in Dorsetshire established a Seal-fishery in Venison Harbour, and his two sons occupied it for some years after their father's death. It proved a very good one, but their nets were extremely injured every season, because their people did not know how to fix them properly; the consequences were, that those at the two entrances were washed to pieces by the run upon the points, and the others, which extended from one Island to the other, were frozen in the Ice, which formed in the bights long before the season was over. By my method, the nets are not likely to take much damage, and the experience of one Season would most probably enable one to make such improvements as might be found necessary.

That part of Venison Island which forms one side of the Harbour has an even surface covered with short Heath, and it rises gradually, to the South East, for some distance, and there more abruptly, as it does at each of the points: and there is plenty of room for all the necessary erections.

Stoney Island has a flat shore, but the land rises abruptly, at a short distance above the water's edge, to a very considerable height, but there is room enough to erect a long Shed to keep the Boats under in the summer time, which would preserve them from the weather and make them last much longer. Other necessary buildings might be erected there also. Perhaps the Oil & Skinning house and the vats would be better upon that than on the other island. For one reason they would be best upon Venison Island and that is, that they would be safe from an Enemy, in time of war; as I could at a trifling expense prevent an Enemy from entering the Harbour, or landing upon Venison Island; but a Ship's Crew might always land upon Stoney Island, and destroy whatever might be upon it. The Boats, however, might be saved by taking them over to Venison Island. Stoney Island is so high, rocky and uneven that it would be a work of very great labour, difficulty and time for an Enemy to bring Cannon to fire upon Venison Island, and they would make use of small-arms to a very great disadvantage

– 1800 feet of 1 ½ Plank, 6000 feet of inch board & 4000 feet of ½ inch board would be sufficient for the Plan on the other side. [pg. 19]

Plan of a Dwelling-house for a Sealing-crew. Scale 10 feet to an Inch.

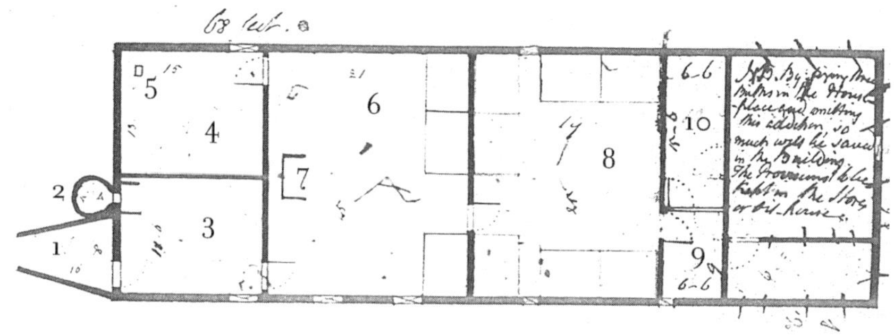

Explanation of the Plan

1. Breakweather, 10 feet by 8. Door-way 3 feet wide & Roof to slope down to 6 feet.
2. Oven 4 feet diameter, with a flue of wattles & clay before the mouth of it.
3. Entrance-room, in which firewood will be staved, to supply the Daily consumption 15x11-6.
4. Master's Bed-room, 15x13.
5. Close Stove to warm it, 1x8 in.
6. Sitting-room and Kitchen, or House-place for the Crew, 25x21, with 3 Bed-berths.
7. An iron Cabouse for cooking *vide* Page 15.

8. Bed-room for the Crew 19x25, with Berths each 6x5 and will hold two men. They are marked in red ink.

9. Beer Cellar 9x6-6.

10. Cellar for Spirits, 15-6 x 6-6.

The outer walls to be 8 inches thick, & Partitions six inches. Made of studs and lined with ½ in. board. The height 7 feet under the Beams. The Roof covered with inch boards, pitched paper over them and a six inch coat of sods over that. The fall from the Ridge to the Eaves only two inches.

The House to be elevated four feet above the ground, to prevent being drifted over. A double floor, with a space of eight inches between, and that must be filled up with Sand or Peat soil. The Doors 3 feet wide, and the dotted lines shew which way they are to open.

The windows to be near three feet wide, excepting those in the Men's Bed-room, which are only two.

The Break-weather has no Door and is to be entered by a Step-ladder. The Smoke of the Cabouse to be let out by a copper funnel.

N.B. The same sized house will suit any strong Crew of Men.

[pg. 20] Querie. Would not Seals Hearts and Tongues make good provisions for Newfoundland Planters, and the Slaves in the West Indies, if cured in Pickle? I do not think that the Hearts would imbibe much salt. They are very good when eaten fresh.

Explanation of the Plans on the other side

Oil House

1. Oil House – 53 feet long, 30 feet wide, and 10 feet high. It must be elevated two feet above the surface of the ground, and the corner under the Stairs cross two inches lower than the opposite diagonal one, that all leakage of oil may run down to that corner, where there must be a plug-hole made through the floor to let it run into a tub, set under the house to receive it. The outer wall may be built with studs, chinced [*sic*] with moss, and lined with half-inch boards. Inch boards may be nailed on the outside to overlap each other, with the studs chinced with moss on the

outside also. The floor to be laid with 1½ inch Plank, neatly jointed, the seams covered with two slips of brown paper, Pitched. The Wash-board all round must be of 1 ½ inch Plank papered and covered with inch board also.

2. A Pen for Salt, 23 by 7 feet, and six feet high.

3. A Vat for Pickling the Seal-skins in, 25 by 15 feet, and 6 feet high. That must be made in the same manner as the floor of the oil house, and a Pug-hole in the lower corner to let off the Pickle.

4. A Stair-case up to the Net Chamber, 2 ½ inches wide each flight.

Net and Skinning Chamber

5. The Net-room 60 by 30 feet and ten feet high with some Beams low across at six feet from the floor to make an open loft to put seals up there, that they may be thawed by the heat of the stove, or they cannot be skinned in the winter.

6. The Skinning Vat. The Bottom & sides to be made in the same manner as the floor of the Oil-house, & a Plug-hole through the end wall, near the stairs, for the oil to be conveyed into the Washing-Vat by means of a spout. That Vat to be 25 by 15 feet, and one foot high, with inch boards laid on their edges at two inches distance to skin the seals upon.

7. The Landing place at the head of the stairs.

8. A Trap Door 4 by 4 feet to hoist nets and seals through.

9. A hole cut through the floor to drop the Seal skins into the Vat below. It is to be one foot square.

10. A large Carson stove to warm the room.

11. A Door, with a broad spout placed underneath to launch the Ran-fat into the Rendering Vat.

N.B. The Oil house must have two windows in it which were forgotten in the Plan. The Stair case to have a Door both at the bottom and also at the top. [pg. 21]

Plan of an Oil-House with a Chamber over it for the nets and to skin Seals in

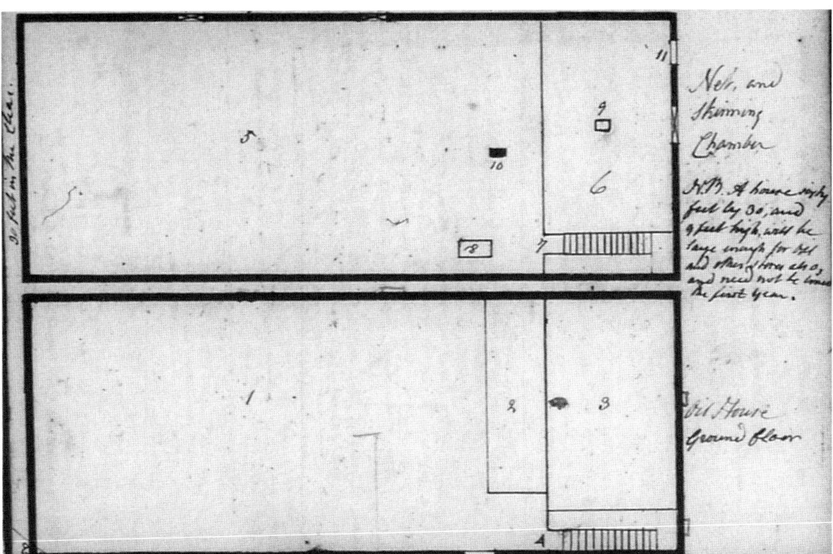

[pg. 22] Care must be taken to find a good foundation for the Posts which support the Sleepers of the Oil house, as the weight will be extremely great after a successful Sealing season. If a natural one cannot be had, which is very unlikely to happen, recourse must be had to squared, large logs laid horizontally. The Beams for the Chamber and those for the loft need not be stronger than usual, as they may be supported by Posts placed under them. Hanging Poles will be wanted, to dry and mend nets upon.

All Glass windows for Labrador should be double glazed, to keep out the frost, and also to prevent their being covered with ice during the winter, which is occasioned, I suppose, by the freezing of people's breath and the steam of hot water.

The Oil house must have a Break-weather before the Door and also a spring, Green Door, or the men will scarce be able to work in it before March. They must skin in worsted gloves, or they will make but very slow work of it.

Pumps must be provided at every Sealing post, to put down occasionally (as they must not stand in the water during the winter) to supply the vats with water, or the carrying of it will be a very labourious work. When

I was in Labrador formerly, I invented a square, and a bent one, to fit the rounding of a Shalloway's side, and it answered as well as any other. It was made of 1 ½ Plank.

A large screw-press is also necessary to keep the dregs of the fat in; and I am of opinion that a pair of iron rollers would answer admirably, and soon repay the cost of it. The loss both in the quantity and quality of the oil, by the usual methods of skinning and rendering which have invariably been practiced hitherto, is incalculable. I cannot suppose it to amount to less than fourteen per Centum, in oil only, and the nets to take ten per Centum Damage, more than either would do by my methods, as well as not kill half the number of Seals.

Where the land at a Sealing-post is nearly upon a level, as is the case at Cape Charles, the net and Skinning Chamber should be built upon the Oil-house as directed in the Plan, and a small window will be necessary upon the Stair-case: but where there is a considerable rise of ground, as there is upon Venison Island, they had best be built separate. The Oil-house near to the water side, for the convenience of shipping the oil, and the Skinning-house at the back of the Rendering Vat that the floor of it may be higher than the top of the latter, for the Ran-fat to be slid down into it. As to the skins, those who drag the seals up to the skinning-house can drag them down to the Oil-house on their return. [pg. 23]

To estimate the quantity of oil which any number of Seals will produce

When I was in Labrador I had a large old Harp-seal skinned at my own house, and the oil rendered with the greatest care, and measured by a wine measure, and it produced eleven Gallons. As the generality of Harps are not so large, and Bedlamers (their young before they show the Harp upon their backs) are much less, and their numbers nearly as great, I estimate the quantity of oil, which each seal will produce one with another, at seven gallons. At that estimate thirty six seals will produce one Tun, and a thousand will amount to exactly Twenty eight Tuns if there is no wastage. [pg. 24]

An Improved Plan of fixing the Seal-nets in Venison Harbour

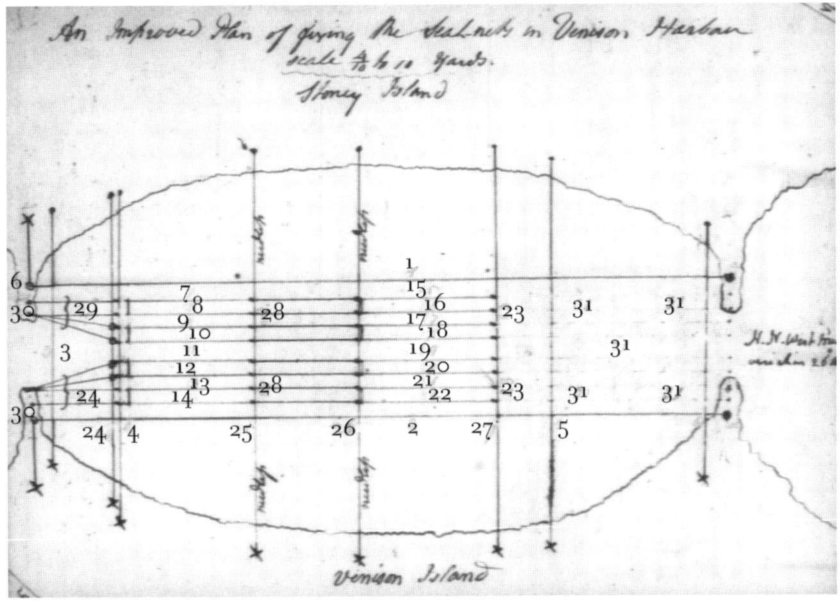

[pg. 25]
Explanation of the Plan on Page 24

By this Plan the Harbour is supposed to be two Hundred yards long and
one hundred broad; but it may be more or less, as I never measured it, nor
have I seen it of twenty four years. But be the dimensions of it what they
may, the nets must be fixed according to the Plan, although there may
be occasions for more or fewer Murderers, according to the width of the
entrances.

1. West Barrier net, 190 yards long. The mooring to be a towline of 3
inches in circumference; made fast at the North end at [blank space]; the
other end rieved through a Block at [blank space], and then brought to the
Capstan at [blank space].

2. The East Barrier. Same length as the former; the mooring the same;
made fast at [blank space], rieved through a Block at [blank space], and
brought to the Capstan at [blank space].

3. The Saveal. It extends from shore to shore, five yards within the Harbour, that it may not take damage from the run of the sea upon the points,
the mooring of two inch rope, made fast upon Stoney Island and brought

to a Capstan upon Venison Island as all the rest of the cross-moorings must be.

4. The Tail-net. 44 yards long; the head fastened to the two Barriers, two yards from each end, and those ends stopped along the Barriers to the Northward, with a stone at each lower Clew. The Mooring two inch rope. Neither of those nets are ever to be lowered.

5. Stopper net. 43 yards long. Mooring two inch rope, it must lay over the West Barrier, and the other end rieved through a Rachit-hoop fixed in one of the upper meshes of the East Barrier, and then brought to its Capstan. The West end of this net must be made fast to the West Barrier, stopped along it for two yards to the Northwards, with a stone at each lower Clew.

6. The Entry net. It is to extend from shore to shore five yards within the entrance of the Harbour. The mooring two inch ropes.

These two nets are to be lowered occasionally to admit the shoals of Seals which will be continually coming into the Harbour, and raised again so soon as they have passed over.

7 to 14. The lower fleet of Murderers, consisting of four nets of 40 yards each in length. Their ends are to be fastened to the mooring at (blank space) & (blank space), and their bights carried to within three yards of the Tail-net, and they are to be stopped to their respective moorings all the way. By this plan there will be four close Pounds, and five open ones; the two next to the Barrier will be five yards wide, the centre one six and all the rest four. Such Seals as run into the close Pounds will mesh in the Murderers, and those which get into the open ones will mesh in the Tail-net, and the Saveal will catch such as may through or under any of the other nets.

15 to 22. Upper fleet of Murderers. Length the same as the foregoing, and put out in the same manner. The open Pounds in this fleet will correspond with those in the other; but that is a matter of the consequence, for [pg. 26] Seals having a space of 33 yards long, by 40 yards wide will play for some time before they enter the lower fleet, and therefore are as likely to enter the close as the open Pounds in it. That delay will also be an advantage, as it will give the Sealers time to keep the lower fleet from being filled too full of Seals at a time.

23. The Places where the ends of the Moorings of the Murderers [space] & [space] are to be made fast to the cross-mooring, and all the marks • denote Buoys; to keep them upon the surface of the water, that the head-

ropes of the nets may not sink under it, and thereby enable the seals to pass over them.

24. A mooring of 2 inch rope, to support the moorings of the Murderers, and these must be four thimbles of wood fixed upon it, for the ends of the four centre moorings to be rieved through, to keep them parallel with each other.

25. A mooring of 2 inch rope to support the outer ends of the Murderers.

26. A mooring of the same kind to support the inner ends of the upper fleet of Murderers.

N.B. Upon reconsidering this matter I am of opinion that both the above moorings may be saved, as the Buoys upon the longitudinal moorings would be very sufficient.

27. A mooring of 2 inch rope; for the standing ends of the Moorings of the Murderers to be made fast to (space) & (space).

[28]. The places where the lower fleet of Murderers are to have their ends fastened.

29 & 29. The Mooring of the Murderers, eight in number. The standing ends of them are to be made fast upon the mooring at (space) & (space), and the other ends brought to posts upon the two points of the Harbour at 30 & 30 after those of the four centre ones have been rieved through the thimbles upon the mooring [24]. As the points at the entrance of the Harbour are six or seven feet, or more, above the surface of the water all those moorings will be kept from pressing upon the heads of the Saveal and Tail net.

30 & 30. The Posts for belaying the moorings of the Murderers to.

[?]. The Grand Pound 35 yards long and 40 yards wide

30 & 31. The Lesser Pound 16 yards long and 40 yards wide.

N.B. I think it would be better to fix the Stopper-net close to the mooring [27?] and have another Stopper-net at the distance of twenty yards before it; as that would make more room to receive fresh shoals of seals in, and keep those which had entered the Murderers from returning.

By the above plan an immense number of Seals would more certainly be caught, as they would have no room to move in when once they had entered the Murderers; and as all the nets, excepting the ends of the Saveal and Entry nets, would be [pg. 27] clear of the dead-water on each side of the Harbour, which freezes a considerable time before the centre part, they would take very little damage from ice, and would easily be taken up at the end of the Season, which is about the 20th of December at that place.

All the nets must be full six feet deeper than the water at the strong tides, to allow for Inching up when the Seals mesh in them, or their foot ropes would be raised off the ground, and many would pass under them.

The foot-ropes must be well leaded, and billets with holes through them would be a good way; as sheet lead will often times come off.

Buoys, of dry fir, should be kept in readiness to fix upon the head-ropes, in case they were found to be wanted.

Having had a considerable deal of experience in Sealing, and seen the different ways hitherto practiced, I am most decidedly of opinion, that this is the best method that has been thought of.

All the moorings must be made of new hemp; and the twine for the nets of that sort called Stopper-twine, which is made with five strands instead of three.

Calculation of the quantity of Cordage & Twine etc.

300 Fathoms of Towline
750 Fathoms of 2 inch rope
1 Ton 4 Hundred and a half of Stopper Twine
1½ Cwt [hundredweight] Lead

This Post will require 17 men (3 or 4 of whom must be Coopers), one large shallop, and 7 Sealing-skiffs. [pg. 28]

Plan drawing of net placement between Cabareta Island and Seal Island (Cape Charles)

N.B. As the Saveal will be 140 yards distant from the Entry-net, in case Seal Island should not be so long, and I believe it is not, the addition of length may be obtained by fixing the Saveal and Tail nets below Seal Island (which is about forty yards broad and the Tickle on the South side of it not more than three, and bringing their moorings to a Capstan placed upon the Continent. Cabareta Island is long enough for the standing ends of those moorings being fixed opposite to their respective Capstans.

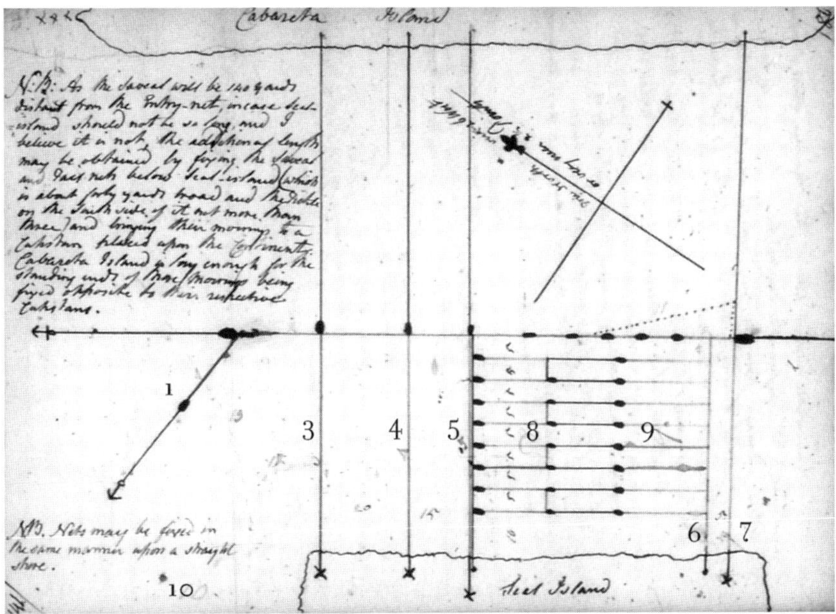

N.B. Nets may be fixed in the same manner upon a straight shore. [pg. 29]

Explanation of the annexed Plan, which is drawn for Seal-island near Cape Charles, and which will also serve for any Sealing Post that is situated either in a wide Tickle, or upon a Straight Shore

N.B. As the length of Seal-island, and the width of the Tickle between it and Cabareta-island are not known, the length of the nets and the exact distances which they should stand from each other cannot be ascertained with precision. The figures in red-ink are to distinguish the Nets; those in black-ink the distances in yards between them; the red-ink lines mark the head-ropes of the nets, and the black-ink lines the moorings; and which are continued under the red lines. The marks • are Buoys, and the x Capstans. The opposite end of each cross mooring is made fast upon Cabareta Island.

1. The Sheer-net, to hem such Seals as attempt to strike across out of Mahar's Cove and go round the west end of Cabareta Island, which they

often do. The dotted line in red denotes the Rout of the Seals in their migration to the Southward.

2. The Barrier net stopped upon a mooring of Shallops Road; laid down with two anchors of 1 C [hundredweight] to the North westward, and ½ C to the North eastward, with a large Buoy where the clews of the nets is fastened.

3. The Entry Net.

4. The first Stopper

5. The second Stopper

6. The Tail net

7. The Saveal

8. The upper Fleet of Murderers

9. The lower Fleet of Murderers

N.B. Another stopper-net may be extended across the head of the lower fleet of Murderers if found necessary. The Stopper No. 5 should be placed behind the Buoys of the upper fleet, instead of before them. This Post will require 17 men and 7 Boats. A large Shallop will also be necessary to fetch the Firewood which may be cut in St. Lewis Bay or Charles Reach; 3 or 4 of the men must be coopers.

10. Part of Mahar's Cove, into which the Seals always go when they come out of Charles Harbour, and proceed from thence by one or other of the routs marked with dotted lines in red-ink.

N.B. Five of the Pounds in the upper Fleet are open, and in case it is found necessary, a net may be fixed on the outside of the East end of the Barrier, as represented by the dotted line in red-ink, lest a great pressure of seals should break through there.

This Plan and that for Venison Harbour will serve for every Sealing post whatever.

All the moorings are to be made of 2 inch rope, excepting that for the Barrier net.

A Bridge must be laid over the Tickle between Seal Island and the Continent. [pg. 30]

Contents of a Rendering-Vat

A Rendering-Vat 100 feet long; 6 feet wide at the bottom; 10 feet wide at the top and 8 feet high will hold about 155 Tuns of Ran-fat. The receiver must be 11 feet wide; and 107 feet long. A washing vat 10 ft. long, 6 feet wide and 5 feet high will be large enough for the above.

Explanation of the Plans on the other side

1. Net-house 20 x 15 x 9 with a Door of 3 feet in front, a window of 3 feet opposite to the Door, and a Carson Stove in the centre.
2. Salt-pen 12 x 15-6; being part of the Salting-house, which is to be 9 feet high.
3. A passage between the Salt-pen and Salting-vat for the Seal-skins.
4. A Door and window opposite to each other in the centre; each 3 feet wide.
5. The Salting-vat 12 x 15-6; being the other part of the Salting-house.
N.B. If that building is attached to the Skinning-house, as in the Plan, there must be a Pickling hole of 2 feet square in the centre of the top part of the partition wall, or there need be no partition wall between them.
6. That part of the Skinning-house appropriated to thawing the seals.
7. The Skinning-vat, 15 x 15 x 1, which occupies the remainder of the Skinning-house. The bottom of it must be raised so high above the floor as to admit what oil may run out in skinning, to be conveyed by a rock into a Gulley [location on diagram is uncertain].
8. There must be a Door of 2 ½ feet in the centre, to enter by, and one of two feet at the end to convey the ran-fat, in Gullies, under the Gallows, to be hoisted up and shot into the Rendering-vat by means of a spout; if the latter cannot be built close to the end of the Skinning-house.
9. The red line denotes the Beams of the Gallows, for hoisting the fat up to the top of the Rendering vat, and the two round black spots are to represent the Posts of the Gallows.

The red line marks the outer edge of a Platform 3 feet wide, to be built within three feet of the top of the Rendering-vat, for men to walk along. The Receiver of the Rendering vat 107 x 11 x 2.
10. The sloping way up to the Platform six feet wide, and as long as will make the ascent easy.

N.B. Where the ground will admit of it, those Buildings may join to each other, to save partition walls; otherwise they may be built separate. None of them need have any lining the first year, if time or materials fall short;

but the Skinning and Net houses should be lined the second year. A Carson ſtove muſt be fixed in the Skinning-house, or the skinners cannot work nor the seals be thawed. These houses are intended to be built upon the ground.

11. Porch 8 feet wide and open entrance 2 feet wide, to keep off the drifted snow, and ſtow firewood in, to supply the ſtoves.

The Roof to fall each way, and slope two inches. [pg. 31]

Plans of a Skinning-house, Salting-house, Net-house and Rendering Vat

N.B. If the ground will not admit of the Skinning-house being built at the diſtance of six feet from the Rendering-vat, a spout, two feet wide with sides 18 inches high, muſt run from the Gallows to the top of the centre of it.

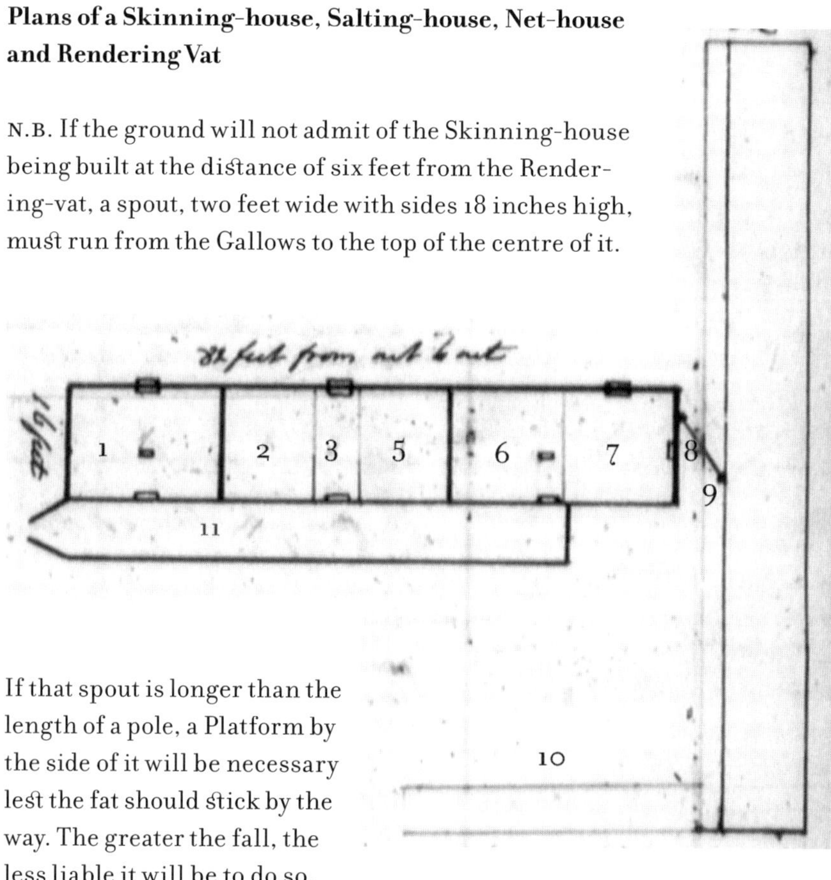

If that spout is longer than the length of a pole, a Platform by the side of it will be necessary leſt the fat should ſtick by the way. The greater the fall, the less liable it will be to do so.

8400 feet of 1 ½ Plank, 6000 feet inch board; & 2500 feet of ½ inch boards will be about sufficient to compleat this Plan.

4000 feet 1 ½ Plank, 3600 feet inch board & 3600 feet of half inch will be sufficient for a Store & Oil house of 60 x 30 x 9. [pg. 32]

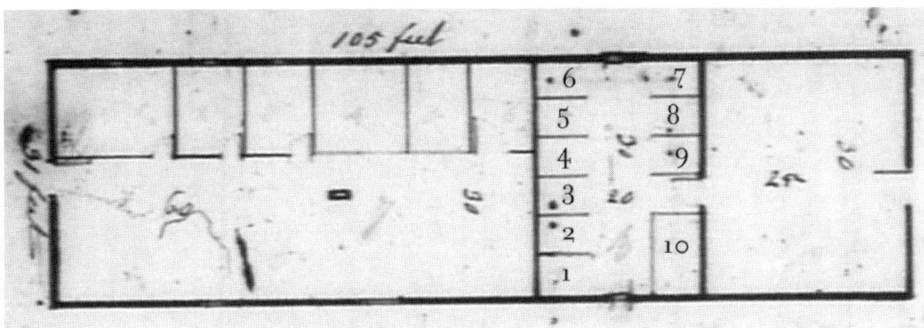

A Plan of a Storehouse, Stable and Hay-room

Explanation

The Storehouse 60 feet by 30, and [?] feet high

1. A shop for keeping an assortment of Hops in is 15 x 12

2. A room for such thing as require to be locked up 7 ½ x 12

3. A room for the same purpose 7 ½ x 12. Doors of both 2 [?].

4. A Pen for Maize 12 x 12 with sliding boards in front.

5. A Pen for Bread 8 x 12, with sliding boards in front.

6. A Spirit Store 9 x 12 with a Door to Lock.

7. The open part of the Store 60 x 18 with two windows 4 ft. wide, and the Door of it and No. 6 4 feet wide.

8. The Door of No. 1 2/6 [feet/inches] and windows 3 feet.

B. The Stables for 6 Cows, a Bull, 2 Horses & Poultry.

Standings 5 x 6 for six Cows; divided by Rails

Standings for 2 Horses.

9. Standing for a Bull.

10. A Pen for Poultry, partitioned off with Battens to admit the warmth occasioned by the other animals.

The whole of the Stable 20 x 30 Door 3 ft. & windows 3 ft.

A Room for Hay 25 x 30. Door 4 feet wide.

The outer walls to be studded; chinced with moss and inch boards to overlap each other on the outside, and the inside lined with half-inch boards edge to edge.

The Roof to have only two inches fall; covered with inch boards; papered and half inch boards over the paper, with sods over all.

The floor of the Store & Stable 1½ inch Plank; of the Hay-room longer.
  Will require
     5600 feet 1½ inch Plank
     7000 feet inch Boards
     6500 feet half inch boards
     A Carson Stove, to dry the Store
     2 Bundles Brown Paper [4 lb]
     2 Barrels Pitch
     1 ditto Tar
     with 4½, 3 & 1½ inch Nails. 4 Locks & 7 windows

Total quantity of Plank & Board wanted to compleat all the Plans and a
Dwelling-house

1½ inch Plank	1 inch Board	½ inch Board
23,000	52,000	45,000

[pg. 33]

Plan of a Merchant's House for Labrador

Explanation

N.B. The House is to [be] elevated upon Posts and Shores, five feet above
the ground, to keep it dry and prevent the windows being filled up with
snow. The outer walls are to be seven inches and a half thick, and the
partitions seven inches. They are to be built of studs flattened to six
inches one way, but left of their original thickness. The outer walls are
to be chinced with moss on the outside, then covered with a course of
brown paper with Pitch and Tar; and a course of inch boards laid over the
paper and overlap each other. The inside of the outer walls, as well as both
sides of the partitions to be chinced with moss and a course of half-inch
boards jointed and laid edge to edge. The floor to be of two course of inch
boards, jointed, four inches assunder and the space between filled with
moss, sand, or [Peat] earth. The Rafters to be made of 2 inch Plank laid
edge way, at the distance of a board's breadth assunder. The Roof to be
a course of inch boards jointed; laid up and down, and their edges meet
over the centre of the Rafters. A course of brown paper laid over the whole
or only over the seams, and paid over all. A course of half-inch boards
jointed, laid over that with their edges meeting over the centre of the inch

boards, and a coat of sods, six inches thick laid over all, but those boards muſt be paid with Pitch or Tar. The Ridge of the Roof to be over the North Partition, and have four inches fall. The two outer Doors to be three feet wide, but all the reſt only two feet six inches. All the windows to be sashes to draw up, excepting the skylights. The large ones to be of four squares in width & three in height, each sash and the small ones two in width. The Shutters to draw up with Rollers, the same as the Sashes, and all the windows to be Double glazed, and the upper squares of the Skylight to be of thick Naval glass.

The red figures denote the number of each room, those in black the dimensions lengthway of the House:

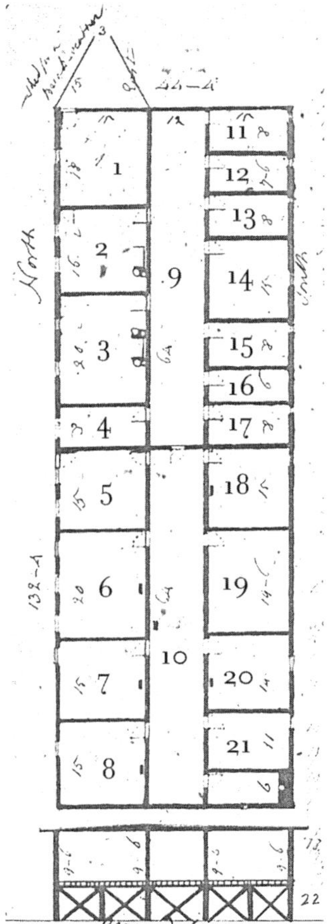

1. Entrance & Wood-room, to supply the fires.

2. Wash-house. Has an iron Cabouse & a Copper, with funnell.

3. Kitchen. Has an iron Cabouse with two Coppers.

4. Dairy, with a wire window for summer & a glass one for winter.

5. Hall for summer entrance, but made up in winter.

6. Dining room, warmed by a Carson ſtove.

7. Merchants Bed-room, warmed by a Carson ſtove.

8. Spare room with a Carson ſtove.

9. Gallery to the servants' end of the House.

10. Gallery to the weſt end with a green Door between them, and warmed by a Carson ſtove, and two sky-lights of 4 feet square fixed over each Gallery.

11. Woman's room

12. Spare-room

13. Cook's ſtore

14. Woman's room

15. Beer Cellar

16. Woman's room

17. Surgeon's room with a Carson stove

18. Wine and Spirit Cellar, lined two inches & chinced

19. Clerk's room, with a stove

20. Pantry

21. Privy, or water closet

The breadth of the Rafters forgotten in the elevation. They are to be 8 inches at top & four at bottom.

N.B. All the Women servants must be the wives of the House-hold serv-ants, or there will be no peace in the house until they get husbands, and may probably [ink stain] those who are not kept in the house: for all women in that Country are strict absolutely of the first great Command-ment. Increase and Multiply. Those who have been married some years and have no children should be preferred to young ones. [pg. 34]

Smoke-house & Bake-house

As an oven will be necessary and also a smoke-house they may be built at a very small expense in the following manner.

Build a House of Studs 24 feet long; 15 feet wide; seven feet high at the sides and ends and eleven feet high in the centre. Rafter, Rind, and Sod it and chince between the studs with moss. Build an oven in the centre, and that together with occasional fires will supply a sufficient quantity of Smoke. If a Leanto was added at one end, six feet wide, that could serve for the Woman's Privy.

Several Beams must be fixed across to hang Flesh and Fish upon, and the top of the Gable-ends left open, to carry off the superfluous smoke.

To build a permanent wharf upon that part of the shore which the tide flows over

It must be built of the stoutest trees squared. Lay three of them parallel to each other from high to low water marks. Lay one across the outer ends of the former, and dub the under sides away so much, that the outer side of it shall stand perpendicular and the upper side horizontal. Lay other cross Beams the whole length of the long ones at four feet assunder, and Beams over them again and build up as before. Continue so to do until you have gained the height required, then lay a flooring of Plank, and nail Planks perpendicular all along the two sides and the ends.

Take particular care that the sides and outer end are perfectly perpendicular, or the ice will tear the wharf to pieces in a winter or two.

A Platform, built in the same way as the bed of a Stage, may be run out from the head of the Wharf to such a distance as will admit of a pair of skids being laid from the head of it to any Ship which you want to load or unload. As that will not stand one winter, it must be taken in before the ice makes.

I have seen sunken wharfs made of Beams trunnelled across each other, open at the sides and filled with stones, but they were soon torn to pieces. The only way to build a sunken wharf in water subject to such thick ice and for so many months together as in Labrador is to fix the Beams in Caissons of strong Plank, and nail the Planks to the Beams with strong spikes. Care must also be taken that the Caissons stand perpendicular when sunk to the ground, and that the base of them be greater than their height. They are to be placed at such a distance that a set of Beams will reach from one to the other, and a pair of skids from the outermost one to a Ship laying afloat at low-water with her Cargo in. N.B. The bottom of the Caissons must be built upon a compleat tier of Beams cut away to suit the ground they are to stand upon, and as they will be wanted to be water-tight for a short space of time only pitched paper over the seams of the Planks will do as well as caulking. The lower tier of Beams in the inside must be an entire one also, and the Planks nailed perpendicular. [pg. 35]

The quantity of Plank of 1½ & of Boards to build the Dwelling-house

Planks of 1½ 15 feet long & 10 inches broad — 477
Inch boards, ditto — 1581
½ inch boards, ditto — 1029
N.B. When in Labrador I bought inch boards at Quebec at 20 /p^r. hundred.

Glass

236 squares 15 x 12 @ 2/6 — £ 43-5-".
74 d° 13 x 12 @ 2/2 — £ 7-"-4.

16 d° 12 x 12 @ 2/1 – £ 1-13-4

16 d° 12 x 12 (naval supposed to be @ 6/) – £ 4-16- ".

[Total] – £56-14-8

Other Materials

8 Good brass Locks with hinges & every thing compleat

1 good plain d°

3 Common d°

9 Thumb latches

13 good iron hinges

4 Bolts

10 Good Doors

13 Common d°

3 Green Doors with springs, hinges, etc. at 2 - 12

6 Quires[8] brown paper (if whole sheets, or 12 quires if cut in strips of 2 in.)

2 Barrels of Pitch & 1 d° Tar

Nails and Brads,[9] Filons for Paste. Paint, oil, etc.

If the roof was covered with thin Mill-board, and the inside lined with the same the quantity required would be 3 Tons, and the cost £168, at 56 / p^r Cwt. [pg. 36]

Plan of a Cabouse for Cooking in, with the funnel to it, and part of the roof of the House

Explanation

1. Posts which support the sleepers

2. A Sleeper

3. Bottom floor of the house

4. The Elevators of the upper floor

5. Upper floor of the Kitchen

6. Legs of the Cabouse

7. Bottom of ditto

8. Sides of ditto

9. Two Pot-cranes
10. Top of the Cabouse
11. Funnel
12. Ends of the Rafters
13. Boarded Roof of the House
14. Fenders of Plate-iron, fastened to the funnel above & the bottom ends to ſtand half-way between it and the Rafters to prevent their taking fire.
15. Top of the sod covering
16. The top funnel which is wider at the bottom and is nailed on upon the roof of the house.
17. Top of that Funnel
18. Legs of the Cap
19. Cap
20. Bottom of a tin plate, which is to be occasionally hung upon the top of the Cabouse by hooks and rungs and hang down, to make the fire burn better & prevent smoking.

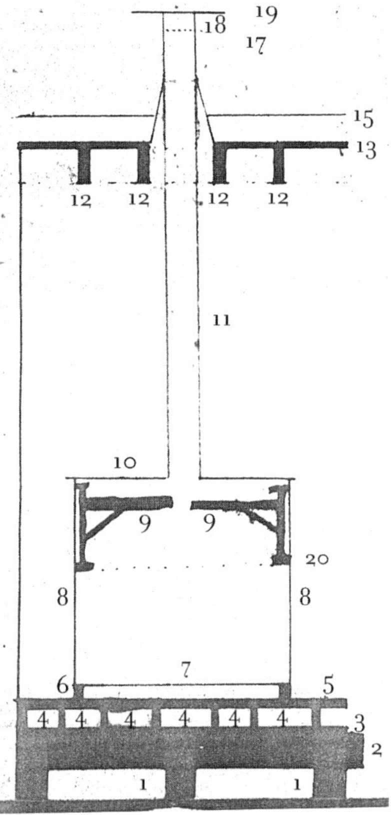

N.B. The sides of the Rafters next to the funnel are to have tin nailed over them, for greater security from Fire, and a sheet of lead laid upon the floor, under the Cabouse.

N.B. The hole in the top had beſt be one foot diameter for the firſt joint of Funnel, which muſt taper to six inches on the top. A Fireplace, whose sides & back are built of Brick or Stone, with an iron top will do equally well. [pg. 37]

Deer Pounds

The quickeſt way of building them will be, to fix ſtrong Poſts into the ground; support them on both sides with shores; saw notches in them wide and deep enough for the ends of the nails (taking the pieces out with a Chizzel) and nailing a bottom over all those notches with one nail above and another below. All the wood may be cut upon or near the spot.

Trap-house for Wolves, Foxes, etc.

Build a house as directed in Page 50 of Labrador Companion. Make the Partition between the traps and the apartment for live-baits of cast-iron bars, or number one wire. Keep tame Rabbits in the apartment and make a crib for Hay against the back side; nine inches above the floor, with battens of inch boards, both underneath and also all along the front, placed an inch and a half asunder, that the Rabbits may always be able to get at the Hay; build a box for Corn, with a narrow slit at the bottom, that it may supply a narrow trough below, and always keep a block of hard-drifted snow in the middle. The Rabbits will then be able to supply themselves with food and drink, in case the weather should prevent the house from being visited for three or four days successively, which will happen sometimes.

Fix up a Gallows near the Door, and Keep a number of Seals' carcasses upon it, which will attract the Ravens, and they the Foxes etc. When you erect the house, always draw a drag up to it, and scent it with whatever experience proves to be the best kind: for severe frost will totally destroy some, as I have found by oil of Aniseed. A Deer's Paunch, or piece of a Seal's skin with much fat left upon it, are good drags for severe weather, as they will leave their marks upon the snow, and those animals will run by the eye as well as by the nose upon snow. Whatever animal arrives at the Door of the house will certainly be caught, and every one which crosses the drag, winds the carcasses, or sees the Ravens will go there.

To Keep Flies out of Apartments, and which may possibly keep the stout from Deer, Cows, etc., and Mosquitoes from Skin

Put into an earthen pot half a pound of Cantharides;[10] an ounce & half of gourd seed; of Motherwort, Sassafras, root of the St. John's Wort and spirit of [Anis] half an ounce of each; a quarter of an ounce of orpiment,[11] and a good handful of savine;[12] the whole cut small or reduced to powder.

Close the pot hermetically; sealing the interstices with flour paste. After the contents of the pot have boiled sufficiently, take it from the fire and let it stand twenty four hours in a cool place; then uncover the pot, and with a feather smear the frames of the windows and Doors, both of apartments and stables from which you are desirous of keeping the flies. A single coat is sufficient for the whole season, but if rain should chance to take it off, care must be taken to renew it. The smell of this preparation, which is scarce perceptible to man, is so insupportable to flies, that there is not a single instance of one having entered by a window or Door to which [pg. 38] this liquor has been applied. To keep them away from Horses, it is sufficient to besmear the harness, the girths, or the saddle with this liquor. (Moor's Almanac, improved 1807)

To build a temporary Store of boards to stow Goods in, until such time as a permanent Store can be built

Lay boards flat upon each other until you gain the required height, and lay as many lenths [*sic*] of boards as will give you the required lenth. Place other boards with one end upon the top of the former, and the other ends upon the ground, and close edge to edge, cover those again with other boards, whose centres shall lay over the edges of the under tier, and lay a third tier over all. Lay other boards across one end of the shed, one upon another to the top, then stow in the Goods, and make up the other end in the same manner. By so doing you will have a Store divided in the middle into two apartments.

Improvements on the Marten-trap Page 17 of Labrador Companion

The two killers to be made of oak Pork-barrel staves and the end of the upper one to project beyond the end of the Log. There need not be a spring to it, but a hole bored through the outer end and a weight of three pounds suspended by a piece of twine. A short log of wood, cut upon the spot, will make a very good weight; let one end rest upon the ground behind the trap, or lay the top-end over the end of the top killer, as is done with deathfalls. The staple to confine the Killers may be of wood also, and then no other iron will be wanted but the hook; the triggar and four nails.

Provisions for one Man for a Year

Pork 1 ½ Barrel
Bread 3 ½ Cwt.
Flour 1 ¾ Cwt.
Pease 2 ¼ Bushels
Treacle 9 Gallons (6 Hogsheads Beer at 1 ½ Gallon of Treacle)
Oil 2½ Gallons
Vinegar 6 Gallons
Butter ¼ Cwt.
Salt Fish 2 Cwt. (caught and cured in the Country)

To be provided to sell to the people:

Coffee 10 lb
Tea 4 lb
Black pepper 4 lb
Raw Sugar 12 lb
Tobacco 12 lb [pg. 39]

Ground plan of a Receiver to a Rendering-vat, with the posts for the latter to stand upon

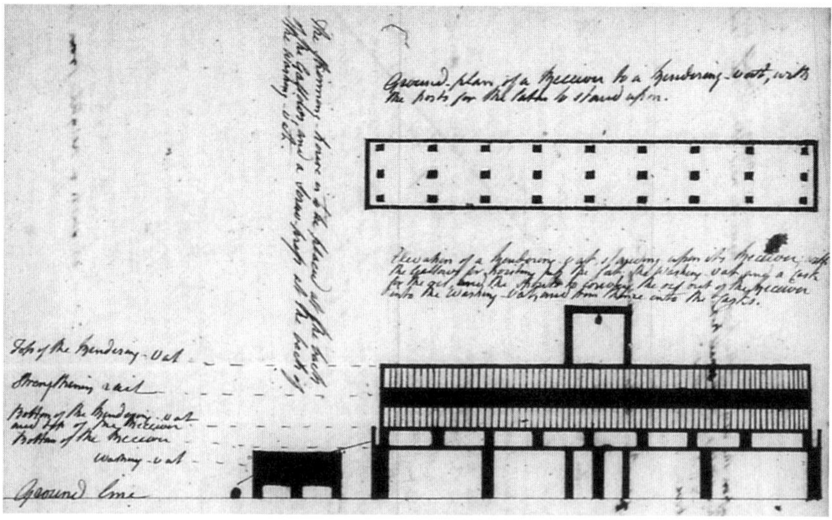

Elevation of a Rendering vat standing upon its Receiver with the Gallows for hoisting up the fat. The Washing vat and a cask for the oil and the spout to convey the oil out of the Receiver into the Washing vat, and from thence into the Casks. The Skimming house is to be placed at the back of the gallows and a screw-press at the back of the washing vat. [pg. 40]

To catch Bears, Wolves, Deer or Foxes upon a hook

A hook for this purpose must be proportioned to the weight of the animal intended to be caught. One made on purpose should have an eye in the top of the shank, to splice a rope or cord in and a ring fixed in the lower part of the shank, opposite the point, to set it by. Find a Tree with a stout branch projecting from it and tie a ring on the under side of it at a foot or more from the trunk of the tree, and another either close to the trunk, or below any other branch. Break the heads off some nails and beat them round. Drive one of them into the trunk of the tree on either the right or left of the side which the branch grows out of, and at such a height from the ground that the animal can but fairly get his chin above it, when standing upon his hind legs. Drive the other [pg. 41] nails higher up, four or five inches above each other, that you may shift the hook according to the quantity of snow which may fall. Rieve the end of the rope through the first mentioned ring, then through the second, and tie a proper weight to it (a log of wood, or a stone) and give such a length of rope that the animal's nose shall nearly reach the ring when the weight rests upon the ground. You are then to hale the weight up, place the ring which is in the lower part of the shank of the hook upon the lowest nail and rest the rope by the side of one of the nails above which will cause the hook to stand properly. Fix a large piece of bait about five or six inches above the hook; a small one halfway between it and the ground; drop other small bits upon the ground to what distance you judge proper; and rub the trunk of the tree all the way below the hook with the same. An animal having found the baits which are upon the ground will be drawn to the foot of the tree, where he will scent those which are upon it, and having easily obtained the lowest, will try for the one above the hook, and in so doing will hook himself under the chin, disengage the hook, and be hoisted up. A loose ring, hung upon the bend of the hook, will do, if you have not one fixed in the shank. An apple is the beſt bait for Deer, and that will catch Bears also. If you use flesh-meat dip it in treacle, for moſt animals are very fond of it. The hook muſt have a beard to it.

An expeditious way of making spring-snare traps for Martens

Cut a length of a ſtraight grained tree, and square it to its proper dimensions; mark the places for the holes, which may be parallelograms; split the log horizontally (the way the holes are to be run), saw out the holes on each side & split out the piece between the cuts; contraċt the holes at a proper diſtance for the threads so as to leave bare room for the breadth of a Marten's head; cut out the slit for the snare with a Lock-saw, or bore it out with a gimblet & burn the edges; bore two holes for the ſtrings & burn them with a nail; Fix in the ſtring and snare; lay the two pieces of log together, ſtop the farther end of the hole with moss, and you may lay the ends of a couple of logs upon the top if you think that necessary. The more holes there are the better leſt you may be kept from visiting it for some Days. [pg. 42]

To destroy Rats and Mice

As those vermin are carried to every part of the World which ships go to, it is necessary to pay ſtriċt attention to them, or they will soon become too numerous for a chance of compleat extirpation. The firſt thing to be done is, upon the arrival of a ship, moor her at as great a diſtance from the shore as you can; fill her Longboat full of fresh-cut green boughs, particularly those of the Birch-tree. At night lash the boat alongside of the ship, lay boards from the gunwale of the ship, or from her loweſt Portholes, to the top of the boughs in the boat; no watch muſt be kept upon the deck, nor any person permitted to ſtir about, but keep the moſt profound silence. The Rats and Mice will not fail to wind the fragrant smell of the boughs, and will feel so ſtrong an inclination to taſte them, that they will walk down the boards and regale themselves. An hour before Daybreak, place one man at the Painter, another at the ſternfaſt and a third at the boards. Take away the boards and vere the boat to a good diſtance aſtern. You are then to tow the Longboat out of the Harbour, toss all the boughs overboard and the vermine along with them, taking particular care that none are left sculking behind. Do the same as long as you find any remaining on board.

Should some get on shore, notwithstanding your utmost care to prevent it, so soon as you perceive that, attack them in the following manner. Mix a small quantity of Treacle with some oat-meal; rub them well together, and add five or six drops of oil of aniseed or carroway. Lay that upon a board and lay it where they can get at it. Feed them in that manner for four or five nights, then give them only half the quantity which they usually eate, and the next night give them more than they can eate of the same food with an addition of Arsenic, or carbonate of barites, very finely pulvarized; or some powdered cantharides, or Nux Vomica finely grated. [pg. 43]

Provisions

As all kinds of Provisions are much dearer in England than in America, it would be best to have them from thence. It would be still more advantageous to have what live Bullocks and Pigs stowage could be found for; there being a number of convenient Islands all along the East Coast of Labrador on which they might be kept and fattened in the summer, and killed about the middle of September. Some breeding sows might be kept all winter; fed upon Seals until June; then turned upon an Island, together with their produce and killed on the setting in of winter. I have done that, and the summer's run compleatly took out what rank taste their flesh might have contracted from the Seals.

To catch Deer in the Winter by whole herds

So soon as the snow begins to fall, which is in October, the Deer begin to draw out of the Woody parts of the Country, and resort to the open barren grounds; and as the Rut comes on them in the first week in October, that may also be another inducement. The flatter the Barrens are, and the more Ling[13] that grows upon them, the greater concourse of Deer will be there; but such places as are visited by an accidental few only, are not worth attending to. Having found a tract of Barrens, which are annually frequented by great numbers of Deer, build a Pound upon it, not less than one hundred yards long, but as much longer as you find it convenient. Build it of Posts, at ten feet asunder, and rails. The two ends must be thirty three feet wide and will require two Posts only at twelve feet distance from each other and each ten feet from the corner Posts. Rail all up with strong rails excepting the centre length at each end, and the Rails for them must be laid upon each other near those openings, and point NW x SE which should be the direction that the Pound should stand; the NW being the prevailing, and severest drifting wind. At a little distance from each [pg. 44] corner of the Pound build a low hut about eight feet long, with the entrance the reverse way to the Pound, and a peep-hole at the other end, and which must always be kept well plugged up with moss, or any other proper thing, when you do not want to use it and a large faggot should be rammed into the Doorway at those times. The chief food of Rein Deer, in the winter, being the *Empetrum nigrum*, or berry-bearing Heath, you must lay in a good stack of it; but as they will eat the wild rye, or any kind of grass, you should be provided with a stock of Hay to fodder them with. Tie the Heath or Hay in small bundles, prick stakes into the ground before the frost sets in to stand better than a foot above the snow, at fifteen or twenty yards distance quite through the Pound and to one hundred yards from each end. Fodder the Deer, by sticking a bundle of the Heath or Hay upon the top of each of those stakes, and when they have once found out that you carry them food, you may be very certain, that they will soon be induced to follow you readily. So soon as they do that, you may certainly catch them in the following manner. Two men must go together, each drawing a sled load of fodder (which is the way that you must always carry it). If the Deer chance to be to leeward of the Pound, approach it to leeward, but take a circuit and draw the Deer to the wind-ward end and then begin to stick fodder upon the stakes. Whilst one man is doing that, the other must go forward, leave his sled at the entrance of the Pound, and secrete himself in that hut which is most to the leeward of the windward end of the Pound. When the fodderer arrives at the other sled, he must take both of them on, and when he arrives at the other end of the Pound he must put in the rails. When the Deer have entered the Pound, the other man must run out of his hut and put the rails in at that end, and the Deer are secured. To be perfectly sure of success, the Deer ought never to be fired at, and if they were called to at feeding times (for they ought to be fed every Day) they would come the readyer but [pg. 45] that must be done sparingly at first, and until you observe them to attend the call. It would also be best not to take all the Deer at any time; as those that were left, would entice fresh-comers. To all Deer-pounds, there should be a narrow killing pound on one side, and then the gun need not be made use of, nor would there be any blood spilt in the catching pound. I never observed that Rein Deer browse upon the bark of trees, as the Deer of Europe do, but they are fond of the young leaves in the spring, and most so of the willow or ozier. There can be no doubt of bringing Rein Deer to fodder, as well as any other kind, but they should not be shot at, nor should the fodderers ever have a dog; there being nothing that they are so afraid of as a Wolf and a Dog. Tame Deer, trained to the sled, would be excellent to draw the forage with. A small drib of maize placed by each stake of Fodder, would prove very enticing to them. Carrots, Turnips or any juicy vegetable would freeze too hard. It would certainly be an excellent plan, to have some tame Deer run loose; but they must be muzzeled, or they would eat all the forage. They would at the least have the effect of causing the wild ones to stand gazing at them, instead of running away, as they would otherwise do for several times at the first; and until they found out that food was left for them. The Deer are most distressed for food when a thaw comes on for a few Days; after the ground is well covered with snow, which generally happens once in three or four years in the middle of winter, but oftener in March and April, for when the frost returns, the snow is a compleat bed of ice, and they find it difficult to procure sufficient subsistence, and many must die for want. I always observed, that they get extremely weak and miserably poor soon after the return of frost, when a free thaw had chanced to happen. But a continued, severe frost never injures them, as their coats are proof against it, and they get plenty of food, by scraping the snow away with their fore feet, and not with their palmated brow-antlers, as is stated by the great Naturalist, the Count de Buffon. I never once saw one make use of its brow-antlers, although I have watched them, with my pocket glass, a great many times; nor did I ever see marks on the snow, of their ever having done so. [pg. 46]

Turnips

Drill them wide enough to plough between, & thin them in the drills with the hand-hoe; draw them before the frosts set in; top and tail them carefully; dry them, and stow them in a Cellar. The Swedish is the best kind;

particularly for Milch Cows. 9000 would be six per Day for six Cows from 1 October to 8 June.

Snow-Bunting

As flocks of snow-buntings make their appearance in Labrador in the Month of May, numbers of them may be caught with strings of Lark-snares; baiting with Hay-seeds. Sandpipers and other Beach-birds may be caught with those snares also. They will also be excellent for Curlews & Grey-Plovers.

To take what salmon is wanted out of a River that is stopped with a set of Racks placed across it

Provide as many Pounds as there are shoots or strong streams, each consisting of two side racks of about sixteen feet in length, and one head-rack of six feet. Place one Pound direct above each stream, and draw three or four staves out of the Rack direct below. The salmon will immediately run into the Pound, & may be taken out with a landing-net or a gill-hook. Always put the staves into the lower rack again when you have done, lest a White Bear should enter and break the Pound. If one Pound full will not suffice, let in another. If the stream is so strong as to create a difficulty in taking out the fish, provide a piece of Canvass of the size of the head-rack; nail it upon two rollers; roll it upon one, then place both the rollers at one end of the head-rack, and unroll the Canvass until it covers it. The same may be done before the lower rack, that you may replace the staves with more ease. The Racks of a Pound should be strong, that they may bear 1½ inch Planks to be laid over it, for the men to walk upon and to lay the fish on. A few links of a trap-chain put round the heads of two adjoining racks, their ends fastened with wire & then let drop, will rest upon the lower bars & confine the bottoms of the heads, that they cannnot slip open, and they may be drawn up again with two gaff-hooks. [pg. 47]

To construct a Pump which shall not freeze

Make it of inch and a half plank and four inches square in the hollow part; two of the sides must project four inches beyond the other two, and other plank nailed across them again, which must project four inches beyond them, and then nail on two other planks upon their ends. You will then have a casing of four inches wide on all the four sides of the Pump, and the hollow parts are to be filled with dry moss or saw-dust. But before the casing is put on, strips of brown paper, pitched and tarred must be laid upon the seams of the pump, and a coat of the same over the whole of the casing, and an outer casing of inch board over the whole, and which must be payed with Pitch and Tar also. The Break must be made to take off, and it is to be taken off whenever the Pump is not used, and a hood of four thicknesses of Blanketing, with an outer covering of leather put over the head of the Pump, and the spout secured in the same manner. As the Pump will not leak, if the work is well done, the moss or saw-dust will always be dry, and stop the frost and it cannot easily penetrate through the Blanketing with a leather covering upon it.

If a Pump is already made, it may be secured from frost by nailing battens across each side, and the casing boards over them. The bottom end of the casing must reach so low down as to dip an inch or two into the water, and be well secured from its penetrating. The Break should be made to work upon an iron crutch, nailed upon the casing, and the spout must be cased also, but that may be done with leather or canvass, instead of boards.

An excellent covering for Houses

Make the rafters of inch boards 12 inches broad on the Roof and 10 on the Wall-plate; fix them at the distance of four inches asunder, nail the thinnest Mill-boards over them; pay the upper side with hot Pitch & Tar; lay a sheet of brown paper upon that paying whilst hot; pay the upper side of that paper, and cover with a coat of sods. As the lower edges of the Rafters will lay horizontally [pg. 48] if you paste sheets of paper under them the ceiling of the room will be flat and you may either whitewash, or paint the paper. You will find a roofing nearly similar to this in page 178 of the Labrador Companion, but this is the best. A roofing of boards is a very unpleasant one in the winter, as the severe frosts cause continual explosions from them, as loud as the reports of pistols, so soon as the heat of the fires abates, and the great number wanted take up a deal of room in a ship, and the freight amounts to a large sum. The Mill-boards cost now

(Feb. 1811) about /6 each; they will each cover 22½ x 17 ½ inches, a great number will stow in a small compass; they will last for ages, and they will occasion no cracking. They may be used to line the insides of rooms with, instead of boards; they will not shrink as boards do, and may be painted of what color you please. They will be equally good to cover the outsides of studded houses, but then they must be well painted with two or three coats. Although they will take more nails than boards, yet the expense will not be so much as those nails will be very short ones. But they may be laid on with glue & then a very few nails will be sufficient. If laid edge to edge & a strip of brown paper pasted over the seams, each board will cover 23 x 18 inches. The prices of these boards is now 56/ per Cwt. and run about 112 to the hundred weight. But they are made of all thicknesses, to the weight of 3 ½ lb each.

To provide good covering for Rubbing-places

Bare a small spot of ground at no great distance from them; fell some bushy-topped spruce trees, and lay all the branches upon that bared spot. In due time all the leaves will fall off, when the boughs are to be removed, and the leaves are to be laid upon the Rubbing-places as they may require them. Dead leaves of spruce or fir make the best coverings.

To prevent a Ship, Boat, Canoe etc. from Leaking

Pay the seams over with black sealing-wax, dissolved in spirits of wine. Use a small painting-brush.

To make a portable raft to carry into the Country to catch Beavers or Otters

Provide eight boards, four feet long; ten inches wide, and a quarter of an inch thick; fifty inch nails; six bladders or seals' paunches and a few yards of seal-twine. In arriving at the place where you want to use it, fell three length of timber eight feet long, and six inches in diameter: rind and flatten them on one side; lay two of them parallel to each other and four feet from [pg. 49] out to out, and the third between them. Nail the boards across them at equal distances; blow up the bladders or paunches, tie one at each corner and the other two in the centre of each side, and the raft will carry two or three people. Two Beams of 14 feet long by 6 inches in diameter with light Longers nailed across at 6 inches asunder & 3 inches wide, covered with seals skin or Eskimeaux buoys at the corners will be preferable & less to carry.

To fix a net round a Beaver-house by men standing on Shore

This will require four men to do it, and it must be performed in the following manner. Provide two poles, of more than sufficient length to reach from the back of the Beaver's House to the distance required beyond it. Lay those poles upon the ground, one on each side of the house, with their ends pointing to the water. Lay the centre of the net opposite to the centre of the house with the leads undermost, and the remainder of each end properly placed for running out. Place the small end of each pole under the net; lift them up; place each upon the shoulder of a man, and the other two men must take up the but ends of them. The poles are then to be shoved out until the net is conveyed to a sufficient distance beyond the house and then let slip off the ends of the poles, when it will be properly placed.* To take it up again, run the poles under the foot-rope and raise that up to the head-rope, and shorten in by degrees: but another pole should be in readiness lest it get entangled upon sticks at the bottom of the water. Provide two light Poles of 10 feet, with a fork at the small end of each.

To cure the bite of Mosquitoes

Do not scratch the parts, but bathe them as soon as possible with spirits of wine. Don Cossacks.

Bathe with Goulard's lotion – for want of that with salt mixed with an equal portion of vinegar. Edw. Dan. Clarke L.L.D, his Journey through Russia. Vol. 1, Page 299.

---

* Take the net up upon them, two men must strip and walk into the water, one on each side of a house, and drop the net at a proper distance beyond it.

Bathe with salmon-pickle, or a strong solution of salt & water. My own application when in Labrador.

Querie. Might not the previous application of some of those things prevent their biting? Such a discovery would be invaluable.[pg. 50]

Deer Pounds

If a Deer-path shall lay parallel to a River or Bay, but at a considerable distance from it, and it should be desirable to erect a Deer-pound near to the shore of such River or Bay, that may be done by cutting a path through the woods in a diagonal direction until it crosses the real path, and fencing the outer side of it: for the Deer will then be insensibly drawn down to the Pound, and enter it as freely as if it was built upon the real path. By so doing much fencing or carriage will be saved as the Deer must either be brought dead from the Pound, or driven down alive through a fenced lane to a killing-pound near the water's side.

Explanation of the Plan upon page 51

1. The outer fence of the Pound 400 yards long including the Hawks; each side of which is five yards. It is to be made of Posts and Rails seven feet high, and the rails at eight inches asunder for the lower half & one foot above.
2. The Hawk entrances, four yards wide at the outer end and just wide enough at the inner end for Deer to pass through.
3. The sheer fences, which are to be run out to the distance of two hundred yards each, made with three rails and need be more than four feet high, but the Posts to be six feet clear of the ground to admit of an additional rail, if that should be found necessary.
4. The Killing-pound, which is to be fixed in the S.E. corner, as the wind is generally in the N.W. quarter. It is to be ten yards long and four yards wide, with one length of rails to take out occasionally at the end next to the Venison-house.
5. The Venison-house: eight yards long and five wide, with a Pitfal trap, for Wolves and Foxes at the farther end.

When any Deer are discovered at feed, two or three men must go on
each side; and drive them in.

The Killing-pound must be made very strong and high; it being there,
that the Deer will make their utmost efforts to escape. [pg. 51]

Plan of a Deer-pound for any extensive piece of ground upon which Deer
feed in the Winter, and particularly for Kinemakoovick

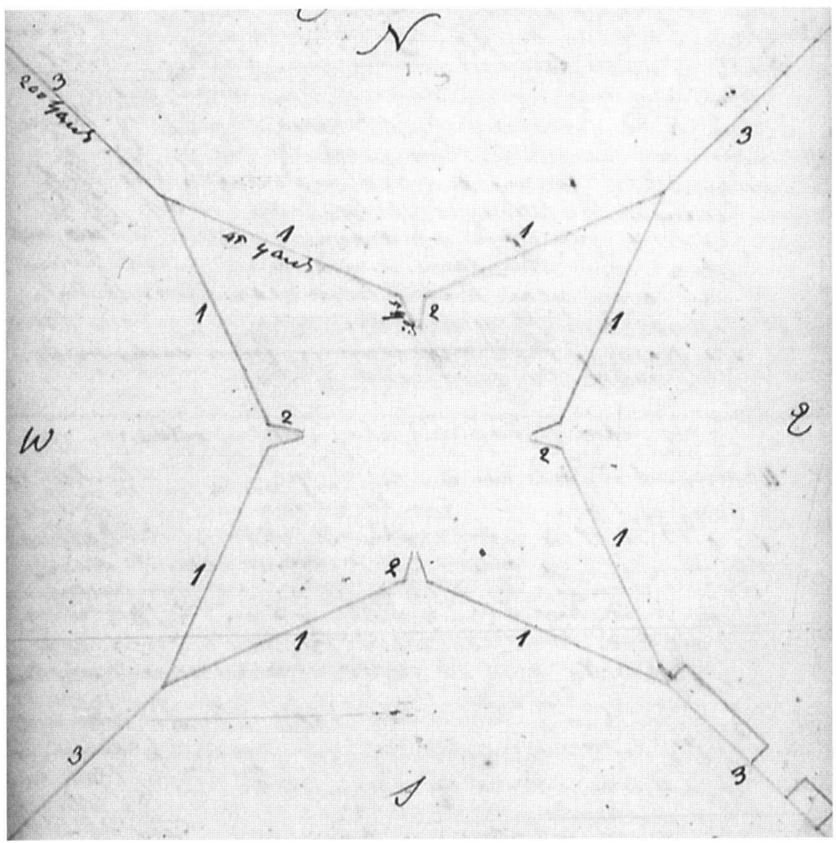

N.B. The Deer may easily be driven out of the Killing-pound into a covered shed, and carried home alive. To build the whole will require 340
Posts, 1520 Rails, 36 studs, 16 Rafters, 10 Beams, roof trees & wallplates &
828 feet of inch boards. [pg. 52]

To catch Deer alive

During the Autumnal migrations of the Deer, they pursue the same routes, and in some few places they leap over a Cliff, when there is a sufficient quantity of snow fallen, to prevent their hurting themselves. When you meet with a place of that kind, which is only to be discovered by observing that they have done so, the whole herd may be taken to a certainty, by fencing a yard underneath the Cliff: because the snow is so light and deep, that they cannot leap up again. I know of two such places upon the same route, which Deer leaped down every year that I lived near them, but I had not then thought of this way of catching them. The season of their migration is, from early in October until the first week in January.

An improved way of Jerking Salmon, Codfish, etc.

Upon a set of Posts, fix a strong Ridge-pole, at the height of seven feet from the ground; at four feet on each side fix a Wall-plate upon another set of posts. Lay Rafters from the one to the other, and a roofing of rinds & sods with a fall of six inches. One foot and a half below the Ridge-pole, fix a set of Poles to hang the fish upon, and mortice them into the Posts. Drive tenterhooks into each side of it at proper distances from each other and those on one side must be opposite to the centre between those on the other. Split the fish, take out all the back-bone, excepting about two inches next to the tail, and break off their heads; score them down to the skin, where they are thick at one inch distance from one score to another, and hang them upon the tenterhooks by their tails, with their backs towards each other. By so doing, they need not be taken in until they are sufficiently dried and the salmon will not get so rusty as if exposed to the Sun. [pg. 53]

An excellent Trap for a Bear, Wolf, or Fox

Open the belly of a Salmon or Codfish sufficiently to take out the guts etc.; fill the cavity with Treacle by the help of a funnel, and sew the opening close up & also the vent. Carry the fish to the place where you intend to commence trapping with its head uppermost, to prevent the Treacle from running out; then tye a string to the tail, score the sides with a knife and drag it after you. Those animals will not only smell the fish, but will also perceive and lick up the Treacle, which will ooze out of its mouth, and follow it to the place which you wish to draw them to. Any bird served in the same manner and plucked, will do as well.

To preserve Bladders, that they may be ready for one when wanted

Blow them up when green, to prove their goodness, then press all the wind out, put them into a keg or wide-mouthed jar, and fill it up with a strong solution of salt in water.

Another way

Cut their necks long; blow them up and tie them at the extremity of their necks; when dry cut off the extreme end of their necks, close under where they were tied, press out the wind, and keep them dry & free from vermin. They must be properly soaked in water before they are used. The vessels which convey the urine from the kidneys into the bladder must be carefully tied.

To catch Birds with fish-hooks

Almost all Birds which swallow their meat whole whether it be grain, flesh, or worms, may be caught with hooks; taking care that the hooks are not too large, and the lines which they are tied upon are fine, supple, and strong. If you bait with Grain, it must be properly softened, by soaking, or boiling. Magpies are so suspicious as not easily to be caught upon a hook.

To destroy Magpies

Make a hole at the thick end of an Egg of sufficient size, take the yoke & white out, mix some powdered arsenic intimately with them and put them into the shell again; put it into a small net, to prevent the ingredients from being spilt, [pg. 54] and suspend them in water, just taken off the fire in a boiling state, for four or five minutes, which will prevent the arsenic from sinking to the bottom, and paste a bit of writing paper over the hole. Make an artificial neſt and put the above Egg into it, together with the shells of four or five more, and the Magpie will not fail to examine them all, and eat the contents of the poisoned one. Carrion Crows will do the same. The Egg should be written upon "A poisoned Egg for Magpies" leſt any person should find and eat it. All poisoned baits for birds had beſt be placed upon the top of a Hovel or other high place, to be out of the reach of Dogs etc.

To anchor an Eskimeau Kyack, to Fish out of

Fix a ring to the Stern, by putting a collar over it and tying that to the rim; Rieve a Saint Peter's line through that ring and fix a conical piece of lead, 5 pounds weight, with five shallop's nails ſtanding two inches out of the sides, to one end of the line, and tie the other to the rim. When the lead is let down, it will anchor the Kyack, and you can easily hale it up again, when you want to move on. You may anchor by the head, in the same manner, if you want to look up the wind or ſtream.

To Keep Foxes in the Summer without expense

If there is an Island in your neighbourhood, of a proper size and at such a diſtance from any other land that they are not likely to swim off to it, turn all your Foxes upon it. Build a house, with a Pitfal trap in the centre & a bridge up to it from the Door, and prevent the Foxes from passing by the side of it. Lay seals' carcasses, or what other food you have, upon the floor of the house at the back of the trap, with a bridge down to it, and keep the trap-Doors bolted. Run drags from the farther sides of the island to the Trap-house until the Foxes have found the food laid there for them. They will walk over the trap to their food and you will have only to unbolt the trap-Doors when you want to catch them. By laying in several carcasses at a time, you will not have much [pg. 55] trouble in supplying them with bait, nor need you run the drag more than a few times until you want to catch them. Some carcasses may be left out side of the door at the firſt, and good, fresh food placed there for a few Days before you intend to catch them, & some of the same sort dragged over the trap & thrown behind it. A Platform may be made from the back of the Trap to the end wall of the house, to lay food upon.

An excellent Sewal for Deer, but more expensive than a feathered one

Stick up stakes at proper distances & extend a sufficiently strong line
from one to the other. Provide a sufficient number of pieces of boards
six inches in length; 1 ½ broad, & ¹/₁₀ thick; suspend those at three feet
distance from each other by a bit of line rieved through a hole bored in
the centre of one end, and paint them black on one side and white on the
other. They will keep incessantly turning with the least breath of wind,
and no Deer will dare to go near them unless forced so to do. The line
should stand about four feet from the ground, & the boards suspended six
inches below it.

The best way to dress Salted Salmon

Lay it in a large quantity of water until the salt is drawn out of it, then boil
it and let it cool. Make a pickle of two parts good vinegar; one of water &
one of the liquor the Salmon was boiled in, but mind to take off the oil.
When it is cold, pour it over the Salmon & there must be more than will
cover the top piece. Add more vinegar, Soy & Keyenne pepper, and what
also you like for sauce, and it will eat little worse than Newcastle Salmon.
If you water it well and boil it, it may be eaten with Egg-sauce, Potatoes or
what else you have.

**An expeditious way of making a Ladder, to rob an Eagle's Eyrie in a Tree
with, or for any other purpose.**

Cut a slender spruce of sufficient size and length, and take both ends off
flat with a saw; nail a bit of a board, two feet long across the ends to prevent the ladder from turning. Nail battens across at one foot asunder. Set
the pole up perpendicular with its top resting against the Eyrie, and tie
it to the trunk of the tree near the bottom. When done with, lean it back
against an adjoining tree until wanted again. [pg. 56]

Fox, or Wolf cubs to catch alive

Having discovered where a litter of Cubs are, dig a hole in the ground if
that is practicable, at no great distance from the place, and make it at least

three feet deep and six feet wide. Lay boards across the sides of the hole
and leave only three feet square open in the middle, and ſtrew earth over
the boards. Drive a ſtake firmly into the ground in the centre of the hole,
and point it. Lay bait round the hole at no great diſtance from it until the
Cubs feed regularly upon it, and then spit a seal's carcass, or some other
fish bait upon the Stake, and throw some into the hole. By attempting to
reach the bait upon the ſtake, they will tumble in, or they will jump down
to get that which is in the hole, and they cannot leap out again. If the hole
is made six feet deep; the old ones cannot leap out, but then the young
ones may not leap in. By chaining a Cub to a ſtake, driven close home into
the ground, and tailing a few traps a little beyond his reach, the old ones
will certainly be caught.

To Tail a large Double-springed Trap upon a broad, flat shore

If you make a slight fence across the shore, down to low-water mark, and
leave an opening in it, rather wider than the trap, and tail the trap in that
opening, properly covered, it will catch any animal which walks along
that shore. But if you intend it for Bears and Wolves only, tail it close to
the upper side, bed it and cover it properly, and fix a piece of bait, either
upon a tree or the point of a ſtake close to the back of it. Should there be
neither trees nor bushes make a slight fence at the back of the trap to
prevent the bait being taken from the back part. Some small pieces of bait
should be dropped upon the ground all the way to the water's edge & the
large piece, that is to be hung up, dragged backwards and forwards firſt.

The Bridge of a double-springed Trap should turn each way; faſten a
piece of salmon, seal's-fat or other good bait upon it; hang the trap to a
tree, with seal-twine, with the bait two feet high, & drop small bits upon
the ground and all animals will be caught by the head. I think that a good
way to catch Foxes with small double-springed traps in the Winter. Tye
them with salmon-twine, one foot high. [?] if a Tree, drive two ſtrong
ſtakes into the ground. [pg. 57]

A Cage-trap for an Otter, upon a Path

Make the top & bottom of boards, and the sides of ſtrong wire, one inch
asunder, with four pairs of Hawk-doors. The trap should be at leaſt twelve

feet long and rather wider than the path. Each side may be made in four pieces; the wires fixed in frames of wood, with one board underneath, and the other upon the top. One pair of Hawk-doors are to be fixed at each end and the other two pairs two feet within the outer pair. The centre part will be the place in which the otters will be confined, and there muſt be a door on the side, to bolt them into bags or nets – or two or three of the wires may be fixed so that they may be drawn out. I advise that the bottom board be laid down firſt, covered with earth, and left for some time, that the otters shall pass over it, that the sides are fixed next & left until otters have passed through, then lay on the top board and after you perceive them to pass through, fix in the Hawks. It would also be adviseable to ſtake the path, on each side, for some diſtance from each end, to prevent their avoiding the trap.

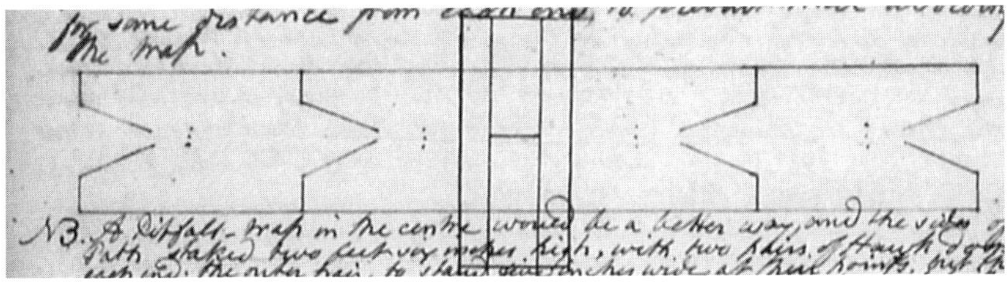

N.B. A Pitfall-trap in the centre would be a better way, and the sides of the Path ſtaked two feet six inches high, with two pairs of Hawk-doors at each end; the outer pair to ſtand five inches wide at their points, but the inner ones only three. Two or three ſtrong wires to be fixed perpendicular two feet within them.

To catch Magpies or Carrion Crows alive

Make an artificial neſt and put the shells of blown eggs into it, which will attraćt those birds. Lime three or four twigs of eighteen inches in length; and attach each to a separate bit of meat about the size of a Pigeon's egg, with a horse-hair, at the diſtance of five or six inches. Place each limed twig near the neſt upon the points of two ſticks, thruſt into the ground, to keep them off the grass and lay the meat at the length of the hair before it. The bird will be immediately entangled with the twig.

A good way to make the Winter entrance into a House which is built upon a Site exposed to great drifting of Snow

The House should be built six or seven feet above the Ground. Then cut a square hole for the Door through the floor of it and attend by the assistance of a ſtep-ladder & make a trap-door to shut over the hole, larger by six inches every way. [pg. 58]

To poison Carrion Crows, Magpies, Hawks etc.

Cut any kind of flesh into such pieces as they can swallow whole; cut a slit in each, put a little Arsenick therein, and sew it close up. Lay two or three of them upon the top of a Hovel, or other place where no Dog can get at them, and they will kill every bird that eats them. A neſt of egg-shells will attraċt Crows & Magpies. In froſty weath(er), the fat of Pork is better than flesh, as it will not only freeze so much, but the fat of a seal is beſt of all; but that is not easily procured in England.

To fix an Eel-net in a River

Lay two heavy Killicks oposite to each other; a sufficient diſtance asunder, and fix two moorings to each: to the end of one of those moorings a log-buoy muſt be fixed, & the other mooring may be tied to it until the net is put out. In putting the net out, the foot-rope muſt be tied to the mooring which has not the Buoy to it & the head-rope to the Buoy. A ſtone of about 10 lbs muſt be fixed to the clew of the foot-rope, which muſt be leaded, & a smaller ſtone put into the Cod of the Net. The head-rope muſt be corked but not much. If the River is deep, the head-rope muſt be tied at such a diſtance from the Buoy, as will permit the net to lay upon the bottom or no Eels will be caught. Eels will not run well before 25th Auguſt.

Plan and Elevation of a Flat, to shoot Geese, or other Water-fowls that frequent a Shoal which lays at a distance from shore

1. The Plan of the inside of the Flat.

1. The Body width at Bottom 3 feet 2 inches. Ditto at top 3 ft. 6 inches.

2. The Stem, length 2 feet

2. Elevation of the Flat.

1. Length of the Body 10 feet

2. The Bulk-head of the Body. Height 3 ft 4 inches

3. A hole in the Bulk-head to point a Gun through

4. The Stem, Height 1 foot 8 inches: To be covered with boards and Caulked.

5. The Keel. Length 12 feet. Depth 1½ inch.

[pg. 59] The sides must flare out to give more clearing, and the stern have a rake of 4 inches. The Thwarts to be moveable, and placed at such a height, that the oars may work easily upon the Gunwale. The oars to hang upon Pins, and be steered with a Rudder, or an oar. The top of the Body to be covered with shifting boards, and the whole to be painted white, that it may resemble a piece of Ice. It must be kept constantly moored at a proper distance from the shoal by four anchors or Killicks; the two head most moorings to rieve through a ring upon the stem, & then through two holes in the Bulk-head of the body, one on each side of the shooting hole, and the two after ones to rieve through a hole on each side of the stern, where they must be belayed to cleats, that they may be haled in or veered out occasionally to place the Flat in a proper position for shooting. A small Buoy to be fixed upon each Mooring, that they may be recovered again when cast off to fetch the Fowls, as that will save the trouble of weigh-

ing all the Anchors. When it is intended to shoot the Fowls, two men to
be conveyed in a skiff to the Flat, and the Skiff muſt return immediately.
There will be sufficient height for them to sit upon the bottom and all the
top boards muſt be kept on leſt Geese etc, that are upon the wing should
see the men, but before they shoot, one or two of them muſt be taken off
so gently, that no noise shall be made. The foremoſt man muſt then shoot
at them sitting, through the hole in the Bulk-head, and the other be ready
to fire immediately upon their rising, and before they disperse. When the
firſt man is ready to fire, he should give a single whiſtle, that the Geese
may elevate their heads, and then fire at their necks, and not at their
bodies, as he will kill more by so doing, and almoſt all of them die upon
the spot; but several of those which drop from the second shot will be
winged only and give trouble to get. Both Men should have a large Gun for
the firſt shots, and a small one to kill the winged birds. So soon as they
have fired, the skiff, which should lay concealed from the Fowls, should
return to assiſt in getting the crippled birds, and fetch the shooters back.

Geese to shoot upon a flat Shore, with tall wood on it

In several parts of Labrador there are flat shores, which extend to a
considerable diſtance with tall woods growing close to them. When you
observe any shores of that kind to be much frequented by Geese, from
September to the time of their migration, cut a clear path through the
woods parallel to the shore, and about thirty yards from it, with side
paths from it to the edge of the shore, about eighty yards from each
other, & make a blind of boughs at the outer end of all those paths, that
the Geese may not see you. Having discovered Geese sitting upon the
shore, which will be at, or after high water, walk along the Path until you
arrive [pg. 60] at the cross-path, which is within a proper diſtance of
them & turn down to the end of it. So soon as you are ready to shoot, give
a whiſtle, to cause the Geese to erect their heads, and immediately fire
at their necks. A Newfoundland dog is a very necessary companion to a
water-fowl shooter, but he ought to be taught to drop to the Gun, and not
ſtir from thence until he is permitted so to do. He should also be taught
to lay by any thing you lay down for him, and watch it until you fire, and
then take it up and carry it to you; for it often happens, that it is necessary
to creep up to what you want to shoot, and the dog would be apt to show

himself if he followed you. If two men are together, the second must fire immediately upon the rising of the geese, and before they disperse.

Remarks on Beaver-nets

To make them more portable, it will be best to have them only twenty four feet long and leaded with Cast-net bullets, one to a foot. That length will generally do in Ponds before they are frozen over, and when a greater length of net is wanted, one or more may be seamed on to the first. As those bullets run about twenty two to the pound, a net of twenty four feet will require only one Pound and two bullets of them. When a Beaver-house is found in a Brook, it will be best not to set the net, which is to be fixed above the house, right across, as the stream will make it bag so much, that they will not mesh well; but sling a large stone, and tie it to the centre of the foot-rope, by a Lanniard of a foot long, and fix a bobber upon the centre of the head-rope. You may then draw the stone along, by haling upon the foot-rope, until the bobber shows it to be in the centre of the brook, and then carry the two ends of the net down the stream until it stands properly & there make them fast upon the shores. As the net which is placed below the house will bag down the stream, it will not prevent the Beaver from meshing, and therefore may be placed right across. Be sure to cork your nets well & do not make any less than ten feet deep, but twelve is still better for the first Beaver that strikes in will tuck it up considerably. Whenever you can do it, fix a second net at the back of the first.

If you can go again the next Day leave the nets placed before the house, to catch what may have been out at the time, as they will strike into them when they return. [pg. 61]

To catch Beavers alive

Provide a sufficient number of iron wires 3 ft 6 inches long, and point them at one end; also a pair of foreceps to take them out of the house; a number of small-meshed nets, to tie over their heads, & sacks or bag nets to carry them in. Place nets before the house, as before directed, and then walk very quietly out upon the house to the outer end of it, there push down the wires about an inch lower than the edge of the water quite down to the bottom, and not more than two inches asunder. You will find where the angle is by the wires going easily down after they have penetrated the upper part of the house, and be sure to bore well all round the end Leſt there should be two angles. So soon as all the wires are placed, break into the house quickly, seize the oldeſt Beavers by the necks firſt, put a net over the head of each and tie it round the neck, then put them into the sacks. The small ones may be taken out by their tails. If you observe that they ſtill can open their mouths, run a bit of line in & out of the meshes, a little above the nose and tie it faſt, but not too close, as they die for want of being able to breathe freely. When you get them home, they muſt be kept in a ſtrong hutch, with a wire front, and a tin vessel for water to drink. Unless you have the proper pairs, they will not breed, for there is reason to believe, that neither will pair again if it loses it's mate; at leaſt not of a long time, as is the case with Swans.

Buoys for a Barrier Sealing Net

Length 10 feet; breadth at top & bottom, the breadth of one plank; depth on the sides half as much; make them of 1½ inch spruce plank; put plugs in the ends of the same; cover the seams with pitched paper, with an upper covering of inch boards, & sling them with two-inch rope, ſtapled on. Fix a dead-eye in the ſtrap at one end.

To build a neat studded House, to be frost-proof

Cut clean ſtuds that carry their subſtance well up, & that are six inches in diameter at their tops; square one side only; shoulder them with a saw, & leave the shoulders one inch thick; place them on the inside of the sleeper & wall-plate, with their buts downwards, and nail them at each end. Nail inch battens, at two feet diſtance, to ſtand perpendicular. Chince with moss between each ſtud (which muſt be rinded) and then nail on the loweſt lining-board; chince with moss between that and all the ſtuds as hard as possible; then lay on another lining board, edge to edge, chince between that and so go on until you come to the top, & then nail a ſtrap of board upon the lining, but the out-side muſt be covered with inch boards; The loweſt one nailed on firſt, chinced within it and then lay on another, overlapping the upper edge of the former, & so go on to the top. In build-ing the roof, nail [a] plank of 1½ inch flat upon the wallplates & projecting over the studs; slope it half an inch from the inner to the outer edge with a plane, cut the rafters to sit fair upon it, nail each, and [lap] a half inch board, of six inches in breadth, in up the feet of the rafters all the length, & fix pitched paper over them & part of the eaves board; the roof being laid on fix an inch lining board on [pg. 62] the inside which must be cut away to admit the rafters, and chinced inside of it with moss. If the work is well done, the winter's frost and summer's heat will be kept out, and the house look extremely neat.

To shoot Curlews, or grey or golden Plover

Preserve one or two of the birds, or stuff their skins in a standing posture, and fix a bit of strong wire to their legs, to fix them in the ground by. Dig a hole in the ground which is frequented by them, if it is not too rocky or wet; a roof must be made over it, rather higher than the level of the ground; with a hole left on one side to get into it, and a small one in front to shoot through; the preserved birds must be pricked into the ground at a proper distance from the hut; an assistant must walk round to drive the birds towards the hut, & the shooter whistle to them in their own natural note, which will bring them to him & they will drop by the preserved birds. Where the ground is too wet or rocky, a low hut must be built upon it. If the hole for entrance & that for shooting through are opposite to each other, the light will shine through & shy the birds.

Wood pigeons will be attracted by a preserved or stuffed one fixed upon the ground, but they cannot be called in the winter.

To preserve birds that they shall be fit for use twelve months or more

Pick, truss and draw them, then wipe them perfectly dry with a linen cloth; draw the skin up, and tie it close round the neck, and then fill them full of Suet or Hogs-lard just warm enough to run into them. They are then to be dipped into the same several times, until they have a compleat coat of it: that done they are to be stowed in a keg or other convenient vessel, and more suet or lard poured in until the vessel is full.

N.B. The birds ought not to be shot, if they can be procured by a net or any other way that will not penetrate the flesh, as the blood cannot be compleatly dried up in the shot holes.

To preserve a large joint of any Kind of Meat

So soon as the animal is killed it muſt be skinned and not cut up until the flesh is cold and set. A linen cloth is then to be pressed upon it, to suck up all moiſture. It is then to be dipped several times in warm Suet or Lard (suffering it to cool between each dip) until it has a coat of at leaſt half an inch thick all over. Provide a keg or cask, of sufficient size for one joint or more, pour warm suet in until the bottom is covered about half an inch, and give a coat also all round the sides, by rolling [pg. 63] the cask until the suet is set: then lay in the joints of meat, head up the cask, and fill it with warm suet, poured in at the bung hole.
N.B. I do not know that this method has ever been tried, but am ſtrongly of opinion that it will answer, as the external air cannot get to the meat, and therefore putrefaction cannot take place.

To prevent a Bear from breaking Salmon-Racks

Fix a Bird-rattle upon it, to be turned by the wind. If you could make it turn with the water, that would be ſtill better, as it would then never ſtand ſtill; which it otherwise muſt do in calm weather.

To build a House of Boards upon the Ground

The materials being all provided, viz., Bricks & Mortar for the fire-places; Planks of Oak or Larch 10 x 2 inches, for the foundation; others, of spruce, 3 x 2, for sleepers for the floors; Sleepers of ditto for the outer Walls & Partitions to reſt upon of 10 x 2; Wall-plates of ditto 10 x 2, for the Roof to reſt upon; Poſts to support the four corners of the House, 10 x 10; other Poſts 10 x 4 to support the Wall-plates at the ends of the cross Partitions; Stancheons 3 x 2 to support all the reſt of the wall-plates, & to nail the Boards to; Rafters of 2 inches in thickness and 10 x 8 in breadth; inch boards for the Walls, Partitions, Floors and Roof, & brown Paper, Pitch, Tar and Moss; mark out the size of the House, take off the surface earth until you come to solid ground. Upon that lay a foundation of ſtones or gravel & then two thicknesses of 2 inch Oak, Larch or Birch Planks for the fireplaces and build them. Level all the reſt of the Ground with stones, and cover them with fine Gravel or sand. Lay the Planks for the foundations of the outer walls & Partitions. Set up the Posts at the four corners of the House & ends of the cross Partitions, by morticeing them into the sleepers, but the tops of them must be flat, & their inner edges even with the inner edges of the Posts, and nail them on. The Stancheons must be cut flat at both ends, placed under the wall-plates at 18 inches asunder, & nailed on; taking care to fix only one to support each Door-case & Window-frame, & leave the other to be fitted in after those are in their places. Nail on the Rafters with their narrow ends downward & even with the outer edge of the Wall-plates, & their broad ends must rest either upon the nearest longitudinal Partition, & meet in the centre and [pg. 64] there be coupled together, and one must be placed on each side of the cross Partitions & the space between filled full of Moss. All the Boards for the Floors, Walls, Partitions and Roof being planed on one side and edged, let in an half inch board across the lower-ends of the Rafters by cutting so much of the latter away, and nail an inch board upon the Wall-plates & outer-ends of the Rafters, then plaster brown Paper, with Pitch & Tar over that & the former board, then lay on the Roof, edge to edge & cover the Beams with strips of Paper as above; across the former lay a covering of half inch boards, with a whole course of Paper over it & a coat of six-inch sods upon that, & the Roof will be compleat. Fix the Door-cases & Window-frames in their places, and fasten them there with the loose Stancheons. Now begin with the outer Walls, by nailing on the boards on the inside of the House & take care to wedge them close down, one upon the another and fill the vacancy over the Wall-plates with Moss. The partition-walls are next to be done by nailing the boards upon the insides of both the longitudinal ones, and of each alternate Room; by doing which, half of the Rooms will be finished so soon as the Floors are laid, and that ought to have been done before the lining-boards were nailed on that the lowest of them might have overlapped the ends and sides of the floor, but the Wash-boards will do that. In laying down the floors, the hollow spaces between the Sleepers must be rammed full of fine earth or sand. The boards on the outside of the House must now be put on; begin at the bottom & ram the vacant space, between that & the lining-board, with Moss, then nail on the next board & do the same until you finish at the top. If you make a cellar in the House, let the Stanchions for that be eight inches thick; that the heat & cold may be effectively kept out. Having finished the outer Walls, begin with the Rooms, & then compleat the Gallery; taking care to proceed in the same manner as with the outer-walls & then nail on the Wash-boards. The Windows should be double-glazed, at four inches between the glass. Instead of sashes, make a casement in each window, of one breadth of Glass, and line the edges of it with drab-cloth or swanskin flannel, to exclude the wind. The Window-shutters must be placed on the inside of the Rooms & drawn up. The smoke of all the Fireplaces & Stoves is to be conveyed out through copper-funnels. If the House is too long to admit sufficient light through a window, at each end of the Gallery, the deficiency must be supplied by sky-lights, which must be double glazed & the upper glass very thick. As the roof will have only two inches fall, there will be no [pg. 64] occasion for Beams.

If the House is built upon such ground as will admit of being under-drained, begin where the water is to make its exit and dig upwards until you arrive under that part of the scullery where a Pump will stand most conveniently; there dig a well, as deep as the ground will permit, and fix a square frame of Oak, Larch or Birch Plank in it, and let the upper part of it be high enough for the Sleepers of the Floor to rest upon: from thence continue to dig upwards until you get beyond the outer wall, and then make a cross-cut the whole length of the upper side of the House; line the drains with Plank of 6 x 2 inches; let the water into the well, by cutting a notch in the upper side of the frame, & let the waste-water out by another in under side: but remember that the bottom of the upper drain is on a level with the notch in the Well-frame, that the water may not lodge in it, and the bottom of the lower drain low enough to carry off all the waste-water, and the mouth of it covered with sticks. Fix a Pump in the Well, and cover that and the drains with Planks of 1½ inches. A trough must be placed under the spout of the Pump & a Pipe fixed in the bottom of the Trough to convey the wash-water into the lower drain. By so doing the Ground under the House will be kept perfectly dry, and there will be a plentiful supply of water for the greatest part of the Year if not for the whole of it. A studded house may be built in the same manner, & if it is a narrow one the front side may be higher than the back, but a broad one will require a Ridge-pole for the upper ends of the rafters to rest upon. The same plan will do for an elevated house with the addition of a double-floor & the frame supported upon Posts and Shores.

A Bolting-door in a large Pitfall Trap, for large animals

Make the frame five feet high & two feet six inches wide; make up the
bottom end with 1½ inch Plank, eighteen inches high; make the Door of
1½ inch Plank & hang it by the top (on the outside) with hooks & gim-
mals; fasten it with a hasp and staple, at the bottom, with an iron pin,
& fix a line in that to draw it out with when standing upon the top of the
Trap. The Yard, which the animals are to be bolted into need not be more
than eight feet wide, & the length of the Trap, and must be covered over
upon the top, or nearly so. [pg. 66]

To construct a light Canoe, for furring

Length & width at the bottom, in proportion to the number of men it is
intended to carry; Depth 13 inches & the sides to flare out six inches;
Stern to project 3 feet & elevated six inches; Stern to flare out six inches,
& be elevated six inches, with a long narrow Rudder. Knees & Floor-
timbers $^6/_{10}$ x $^3/_{10}$ inch. Bottom & sides to be double, & made of Birch
$^2/_{10}$ thick; seams smeared with thick white-paint, and a coat of the same
between the two plankings; cut limber-holes in the floor-timbers & nail
on bottom-boards of Birch $^2/_{10}$ inch. Work it with Paddles & short poles.
The bottom must over-lap the sides, & a slight Gunwale must be laid over
the top of them.

Deer Pounds

The Hawk-racks should be made in two parts; the lower parts fixed firm
in the Ground & stand 3 feet high, and the upper parts made to stand
upon the under ones & turn upon the points of them, & all should have
fine Spruce boughs tied upon the base of them. The Killing-pounds need
not be more than six feet wide, or six feet high, and covered on the top
with logs.

To secure Houses from Wind

The Door-frames must be of such a breadth as will admit of the Doors
being let into them 1½ inches, that a slip of swanskin may be nailed on,

for the Doors to shut upon. If the Fire smokes, bore a hole through the
Wall, at the back of the Stove & stop it occasionally with a Plug. The Beams
& Rafters must be laid upon the Wall-plates, with their ends cut even with
it, & the spaces between rammed full of Moss & secured on both sides
with boards; a half-inch Board must be fit in upon the ends of the Rafters
& a piece of paper pasted over that & the Wall-plates. The lowest Board of
the Roof must project eight inches over the Walls.

A Floating Bridge across a River

Square logs of Wood of 8 or 10 feet in length & moor them, with either
anchors or Killicks, parallel to each other. Lay two Beams across their
ends, nail a sufficiency of Longers across the Beams & two Planks over
them lengthway. Take all up before the frost sets in, or the ice will carry
them away. [pg. 67]

Rendering Oil by the Sun

If a Rendering-vat be built upon level ground, it will require too much
elevation for the Receiver to be high enough for the Oil to run into the
Washing-vat, and therefore it must be Pumped out of the former into the
latter; consequently, the Receiver-vat must be built higher than usual,
lest it run over in the Night.

**To Launch a Vessel which is built upon a broad, flat Shore encumbered
with large Rocks**

Run out a Platform from the Dock of sufficient strength & breadth, constructed like the Bed of a Cod-Stage, and make the Cradle to run upon that.

To lay Shallops etc. on Shore for the Winter

Never lay a shallop on shore under abrupt, high Land because she will be
covered with so much drifted snow as will do her irreparable damage, but
do it upon the flattest shore that you can find, and Block her up high, that
the drifted snow may have a free passage under her. Small-boats had best

be haled up into the Woods, and a shed covered with Boughs, built over
them.

To catch Otters or Black-ducks upon a Path

Where you discover a Path, used by either Otters or Black-ducks, you may
catch numbers of them, at the same time, by setting a Thief-net upon it.
It must be made of Trall-twine for Otters, or Salmon-twine for Ducks;
the length twenty yards, circumference four feet, and Mesh one inch
square. The Hawks must be two feet long; the mouth of it eight inches in
circumference, and kept open by twine of the same size. Hawk-hedges
should be made at both entrances by pricking small spruce boughs into
the Ground & the neck should be of the same color as the [pg. 68] herbage
which grows on the sides of the Path. If the Path runs under bushes, that
is the best place to set it in, as they will not take so much notice of it there.
Or they may be caught in a sheet-net 21 x 4 feet, mesh 3 inches squares,
laid lengthway upon the Path, with the front end elevated four feet & the
tail end pegged down at the extremity, & two or three feet laying flat upon
the ground. It must be supported with small sticks, at proportionable
heights, all the way from the outer end to where it lays upon the ground,
to prevent its moving with the Wind. When the first Otter or Duck gets
entangled it will soon pull the whole of the net down, and all those, which
are then under it, will be taken. Another net, of the same kind, should be
fixed the reverse way, as there is no knowing which way they will come.
Fix them in places as much sheltered from Wind as possible, and at a
distance from each other. The Net must be either brown, or dark green,
according to the color of the Ground.

Cheap Stoves for Servants & Store-houses

Build the Bottom of two course of bricks laid edgeway & the upper course
to cross the lower one; the open part must be two feet high & eighteen
inches wide in the clear; the sides one brick thick; the top the same;
supported by bars of iron & a hole left for the Funnel; the Door of iron
with a blowing-hole in it, and fitted into an iron frame. Or the top of
iron. [pg. 69]

An excellent bait for Foxes & Wolves

Any quantity of old, black, rotten Cheshire-Cheese; one fourth as much
by weight of the fat of a Seal, Goose, or Duck, and as much Treacle, or
brown sugar, as will make the whole, when pounded together, of the
consistency of Butter; add thereto a few drops of the oil of Rhodium,[14] of
Caraway, or of Aniseeds. Keep it in a tin box, and cut out bits the size of
a nutmeg, as they are wanted. Oil of Salmon is excellent & will cost only
labour to procure.

To stretch Cordage

Lay two Anchors upon the Ground, [at] a sufficient length from each
other; fasten the article to be stretched to one of them, and stretch it
with a double-hackle fastened to the other, and keep it off the Ground by
round logs.

To build a circular Pit-fal Trap, on a place exposed to Drift

Make two frames of inch board, ten feet in diameter. Lay them upon the
Ground one upon the other, tye two ropes across one of them; set up a pair
of Shears to be twelve feet in the clear & fix a ring close under their top.
Rieve a rope through that ring & tye one end of it to the ropes which are
upon the upper frame, & hoist it up. Having provided a sufficient number
of squared studs seven feet long, & shouldered them one inch at each end,
nail four of them upon the bottom frame, at equal distances, then let the
top frame down & nail the top of those studs to it. The top frame will then
be secure & the rest of the studs are to be nailed on in the same manner.
Having made a frame, to support the top of the Trap, lay that upon the
top frame. Each of the falling-Doors must be three feet long and four
feet wide, & meet in the centre, which will make the door-way six feet by
four, and as the back part of each door will be shorter, & consequently
lighter than the fore part, weight must be added underneath, to make it
two pounds heavier. The top of the Trap and the doors are to be made of
inch boards. Under the back end of the doors make a bolting-door, four
feet six inches high & two feet six inches wide, of 1½ inch Plank; make a
frame for it, fix that frame one foot above the ground, and hang the door

by its upper end; the lower end is to be secured by a hasp & staple, & is to
be drawn up by a rope. A Bridge must be built on each side of the trap; five
feet wide on the top, ten feet at the bottom and ten feet six inches long, of
stout longers, flattened on their upper sides, & rough inch boards nailed
[pg. 70] across them. Upon the top of the trap and between the ends of the
bridges and the sides of the falling-doors, platforms must be fixed, and
supported, by round feet, four inches high. The inside of the trap must
be lined with inch boards; the edges of the bolting-door frame defended
with iron hooping, & a board nailed round the top frame and ends of
the timbers of the top on the outside to prevent the Drift from entering
there. A yard is then to be constructed, in the manner described in Page
15, the outer corners of which are to extend to the two bridges, and the
Posts at the inner corners high enough to fix a Gallows upon, to lay Seals'
carcasses on it for bait. By the Bridges being covered with boards, if the
animals venture to go up them, they will not be shy of stepping upon the
trap-doors, and, as they must throw their whole weight forward, when
they tread upon them, they will infallably be caught. The edges of the
lining-boards must be jointed, that they may effectually keep out all
Drift, or the Trap will soon be filled with it. One length of railing of the
yard must be made to draw out, and secured by pegs when in their places.
The yard must be longered over upon the top.

**An expeditious way of fixing Beams for floors & Rafters which are made
of Planks**

Cut an inch Board of the same breadth as the Sleepers, or Wall-plates of
your House are, and then cut it into such lengths as you intend the Beams
for the Floors, or Rafters for the Roof, should be placed from each other.
Fix the end Beam, or Rafter in its place, then lay a length of Board upon
the Sleeper, or Wall-plate & fasten it with a single nail in the centre, you
are then to set up another Beam or Rafter & lay on another Board, and
proceed in that manner until all are fixed.

To prevent Leaks in a Boarded-roofed House

Cut all the covering boards to the same breadth, & long enough to reach
from top to bottom of the Roof. Fix the Rafters at such distances that their

centres shall be the exact breadth of a covering-board from [pg. 71] each
other. Nail on the first covering, lengthway upon the Rafters, with the
edges of each board as close as possible, and direct over the centre of each
Rafter. The upper covering boards are to be nailed on with their edges
meeting over the centres of the under boards, and a length of Seal-twine,
or Rope-yarn placed upon the under boards a little on each side of their
centres, & exactly where the upper boards will be nailed. A coat of paper
with Pitch & Tar is then to be laid on & that covered with sods. Should the
shrinking of the Boards crack the Paper & admit water into the seams of
the upper coverings, the Twine or Rope-yarn will prevent its getting to
the seams of the lower covering. That ought to be painted or Pitched.

To fix the Stancheons of Outer-walls or of partitions

All Stancheons are to be made of Plank of such a thickness as you intend
the vacancy between the boards shall be. Saw those Planks into such
breadths as you judge necessary. Cut them into such lengths as will give
the required height to the Rooms. Fix them in their places in the same
manner both at top & bottom, as directed for Beams and Rafters. Lay
the Wall-plates over them, & drive a nail through that into the head of
each Stancheon. In the first instance, set up one of them at each end of
the House & also at each of the longitudinal partitions (if there are any);
then set up one at each end of each cross partition; one is then to be set up
on one side only of each Door & Window, and the places for those on the
other side of them marked; the spaces between the Doors & Windows in
each room & the cross parti[ti]ons are to be equally divided, so as to leave
as nearly eighteen inches between each Stancheon as may be. Upon the
inner edges of those Stancheons which are fixed on each side of the windows, nail a piece of two-inch Plank for the frames to rest upon, and as
many Stancheons, as are requisite, must be set up under them and others
are also to be fixed between the tops of the frames and the wall-plate. The
Stancheons are then to be fixed for the cross-partitions with a wall-plate
upon them also. As the Rooms will then be of the same height on every
side, to obtain the intended decent for the Roof, saw a Plank with a proper
slope and nail it upon the wall-plate of each cross [pg. 72] partition and
lay a Rafter upon it. An additional Wall-plate is then to be laid upon
the remainder of the longitudinal partitions (or outer wall, as the case

may be), and properly sloped away for the Rafters to bear equally. The Wall-plate at the lower ends of the Rafters must be sloped off in the same manner. The Stancheons may be shouldered at each end & morticed into the Sleepers and Wall-plates, but querie, will not the additional expense of labour amount to more than the cost of the nails, and also occasion an inconvenient loss of time?

To cut a Marten-path from Deer-harbour in St. Lewis's Bay to Hawk-river

If any Merchant resided at the mouth of Hawk-river, and had a Sealing-post upon Seal-island near Cape Charles, it would be a very desirous thing to have a path cut, that he might have an easy communication between them. The distance is about seventy measured miles, it would require eight Furrier's Tilts to be built upon it, have eight Marten-traps (each under a back-tilt to keep off the snow in the winter), two head-traps between each Tilt, and a sufficiency of Otter-traps for the number of rubbing-places. By setting in different gangs of Men at the same time, or as nearly so as circumstances would permit, the whole business would be compleated in a short time and as all the Tilts would be built where access could be had by Water, the Provisions and all other necessaries could be placed in them with great ease. That the whole may be finished before the end of June, at which time the Flies begin to bite most intollerably, it would be necessary for his Ship to arrive in Charles harbour on the 16th of May, as it is generally free from Ice by that Day, and it will be safest to enter it from St. Lewis's Bay and through Enterprise Tickle, lest there should be a bridge of Ice left across the Harbour from the South point of Eyre Island to the opposite point upon the main-land, as frequently is the case. On his arrival there, he is immediately to take possession of Seal Island provided it is then unoccupied. The Men requisite for the first year, and to prepare for establishing business upon a large scale the second, will be 1 Clerk; 2 Boat-builders; 2 Assistants to them; 4 old Furriers; 8 Youngsters under them; 2 Top-Sawyers; two bottom ditto; 1 Squarer; 1 Cooper, and 1 Youngster for a Cook to the Builders' Crew. If Seal-Island is occupied, he may take possession of Battle-harbour, which is a good Sealing [pg. 73] Station. I previously suppose, that his ship has taken in a large quantity of 1½ inch shallops planks, & of boards of one inch, ¾ & half-inch from some Port in America, and proceeded from thence through to Gulph of St. Lawre[n]ce and the Straights of Belle Isle, which is the beſt rout for a ship to take to get early into a Port on the Eaſt Coaſt of Labrador. The next things to be done are, to dispose of the men in the following manner, viz. The Clerk to make an accurate survey & Plan of the Sealing Poſt; ſtretch ropes across, in the form that the nets are to be placed, and take the depth of Water, at every five yards diſtance upon those ropes, at the time of High water, Spring tides, which rise about six feet. The head Boat-builder is immediately to begin upon a shallop of 32 feet Keel & proportionable Beam & depth under it; the Timbers of which may be cut in Charles river. The other Boat-builder, one Furrier, two Youngſters, and four other Men (two of whom to be of the Ship's crew) to put Provisions and all requisite necessaries into the Ship's Long-boat and go to the Head of Deer-harbour, where he is to build the Tilt No. 1 on the South side of the Brook; whilſt he and two assiſtants are at that work, other two muſt try if they cannot take off and cure as many rinds as will cover it, and the Furrier with his two Youngſters are to begin the Path, and cut it clear, and erect the Trap-sheds as they go on; measuring the ground as they proceed, that a trap may be placed at every 220 yards. As that Path will cross the Brook which empties itself into the deep Cove that runs in between Deer Harbour and St. Lewis's Bay, a bridge of two squared Beams muſt be thrown across it, and if there is any danger of those Beams being carried away by floods or Ice, they muſt be sufficiently elevated by placing two logs of six feet in length and nearly as much asunder; parallel to the Brook; two others across them, and so on until a sufficient height is obtained; the same muſt be done on the other side of the Brook & the Beams laid upon the top of those Piles. On arriving within a sufficient diſtance of St. Lewis's Bay, the Path muſt run parallel to it until it extends about a mile above the farthermoſt Point that can be seen from the entrance into St. Lewis's river; before which time three men, who are to be sent into Alexis' river will have marked the Path from thence across to St. Lewis's river, and then the whole of them are to cut on clear through to Alexis' river. On arriving at the place [pg. 74] where No. 2 Tilt is to be built, which will be before No. 1 is finished, the place is to be marked, and also a mark set up by the shore-side, that the Boat-builder may see it, as he rows along the shore, and the same is to be done where No. 3 Tilt is to be erected. Two Furriers & their Youngſters are to be furnished with two skiffs, and a sufficient number of Otter-traps and catch what Otters they can in Charles Harbour and St. Lewis Bay, while the ship remains there, and as those who go up the latter cannot return the same Day, they must skin out their otters so soon as they are dry, & spread them so soon as they return on board again, where at least twenty Boards & 40 Busks[15] must be provided. As the Ship's Carpenter must assist in Building the Shallop, and the Master & Mate make the sails, prepare her rigging, etc., it will not be long before she will be finished and ready for sea. So soon as she is (the Tilts Nos. 1, 2, & 3 being previously furnished with Provisions and necessaries, and one Furrier and his youngsters landed on the South side of Alexis' river, above the low ground which is liable to be flooded in the spring, to which place they may be safely sent in the ships Yawl) she must take in Provisions and necessaries for as many of the other five Tilts as she can stow, together with the Tilt building Crew, two Furriers and their Youngsters and proceed to the head of Gilbert's Inlet, where the Tilt No. 5 is to be built; whilst that is doing, one Furrier & his youngsters are to mark the Path across to Alexis river, opposite to the place where the Tilt No. 4 is to be built on the South shore, and on which a conspicuous mark must have been placed by the Furrier, who was landed there to mark the path across to St. Lewis' river. On their arrival there, should they find that they had deviated too much from the straightest line, they must mark a fresh path back again, and try to cut off the angle. That done they are to begin to cut the path clear. When the Boat builder has finished the Tilt No. 5, he is to run round with the Shallop into Alexis' river and build the Tilt No. 4 upon the South side, leave Provisions etc., and a small skiff and a Canoe there, and then proceed to the long, narrow arm in St. Michael's Bay, where [pg. 75] he is to build the Tilt No. 6; from thence to the wide Cove which is higher up and about half way to the mouth of the River, and there build the Tilt No. 7, and then that of No. 8 on the south side of the river, and where it will be easiest crossed in a Canoe. Whilst those things are doing, the Furriers must mark & cut the Path, from 6 to 5, and then on from [5]. The Builder having finished all the Tilts must proceed to Hawk river, where he will find the ship. During the time that the vessel remains in Charles Harbour, if any residents are found between Cape Charles and Cape St. Lewis the Clerk must get acquainted with them, and try if he can discover what other places are occupied between Cape Charles and the North side of Sandwich Bay; if he finds that Venison Harbour is not occupied, nor likely to be so immediately, there will be no occasion to run risks, otherwise it will be necessary to get possession of that place with all speed. So soon after the 1st of June as is convenient, but not before, the Shallop being finished, the Builder and his assistants are to embark on board the Ship and proceed to Hawk river; on arriving near the entrance into Hawk Bay, between Hawk & Stoney Island, a boat must be sent into Venison Harbour and take possession of it. On arriving at the head of the South arm of Hawk Bay, if a proper place for building the Merchant house be found there, that must be reserved for that purpose; otherwise the head of the North arm must be examined. When the place for that house is determined upon, the most convenient place for a Store-house, oil house, rendering vats, etc. is to be sought out lower down the Bay; a Store-house is to be erected with all speed, the Ship delivered, and then sent to Quebeck for Bread, Flour, Hogshead-packs, hoops, Board, Plank & such other things as may be wanting; taking care that there are at least one thousand Hogsheads for each Sealing-post provided against the end of the next summer, for it will scarce be possible to occupy them sooner. On the arrival of the Shallop, in which one of the Furriers and his youngsters are to come although not ordered before, they are immediately to mark the Path [pg. 76] from Hawk-river to St. Michaels river, and then cut it out clear.

As the Boat-builder, who was employed in erecting the Furriers' Tilts, should have had strict orders to observe at what places the Sawyers could be fixed to advantage, and the head builder would have had time to examine Hawk's Bay before the other arrived, the latter, with the approbations of the Clerk, must determine where all hands (the Furriers excepted) must live for the next winter, and repair to it so soon as other circumstances will permit, but in case a sufficient number of Stocks can be found in Hawk's Bay, they are to Winter there, and as near to the Merchant's house as convenient. In case that place should be at too great a distance from Hawk's river, a ninth Tilt must be built upon the south side of that, for the accommodation of the Furrier who is to reside at St. Michael's river.

In cutting the Paths, those which run along the North side of St. Lewis's Bay, and the South side of St. Michaels Bay may be compleatly finished as the men advance, previously rowing along those shores at a

proper distance to see in what directions they must run to avoid hills &
cut off angles. But as those which run across the Country will be liable
to deviate considerably from a straight line, the Furriers must walk up
the Bay, or down the Brook, as the case may be, until they find the spot
where the other Tilt is, and then mark back until they have corrected their
error. In the first place, the Master of the Ship must give them the proper
bearing (by Compass) from one Tilt to another, and in the second, they
must be furnished with a good pocket Compass, to steer their course by.
When they meet with a Marsh or Barren, they must fix their eyes upon an
object in a direct line on the other side & go as straight to it as the ground
will permit, sticking up barked poles at short distances, to direct them
in drifting weather, and cutting up the brush-wood to make a fair path to
walk on. When they meet with a Pond or Brook, after making a mark on
the shore, one must go to the right & the other to the left until each meets
with a convenient place [pg. 77] to cross, then return & take the nearest.
They must then alter their original course for a sufficient distance to get
into the line again, and on their return must try to cut off the angle that
they made in going to the Pond or Brook. In marking the Paths, they must
be sure to mark all the Trees upon their right hand only, and when they
have occasion to mark a fresh path back (to correct an error) so soon as
they get into their former path they must make a sufficient mark to prevent their going wrong again. By keeping a steady pace and observing the
time, by a watch, they will give a near guess at the distance passed over.

In cutting the Paths clear out, the distances must be measured by a
Perambulator, that a Trap may be placed at every 220 yards, which will be
eight in every measured mile. All those places must be marked upon the
side of the nearest tree as high as a man can reach, with a Cooper's mark-
ing-iron, & numbered in succession to the next Tilt, and the miles must
be marked also there, viz. M 1, M 2 and so on to the end of the walk.

The Sheds must be erected across the Paths, in the form of Back-tilts,
by laying a stout ridge-pole upon two Trees at the height of seven feet or
more, & rearing others of 8 or 9 feet in length from the ground upon that
and close to each-other, & they covered with rinds at any convenient time
afterwards, but they may be left without the rinds for the first winter (if
there is not time to procure them) as they will keep off the snow. Where
the Path crosses a Marsh or Barren, Head-traps are to be placed in lieu

of Marten-traps. As the Paths Nos. 3, 4, 5, 6, & 9 will cross the Autumnal rout of the Deer, some of them may be caught in Slips and particular notice must be taken of the numbers of Paths, and of Deer which use them, that Deer-pounds may be erected upon such as are convenient. So soon as a Deer or Bear is killed, it must be paunched, or it will be spoilt if left until cold. The paunches must be cut open, the contents shook out, but not washed, and then dried in the air, and care taken that the Dogs do not eat them; when dry, they must be kept in the store for bait for the Traps. All Rivers & Brooks (if practicable) must have [pg. 78] Racks fixed across them, to stop the Salmon both for food & baits, and so soon as the Frost sets in a sufficient number of salmon are to be taken out & kept in a frozen state for winter's use, and the Racks taken up.

The Tilts Nos. 2, 4, 6, & 8 are to be considered as Resident ones, and all the rest Temporary ones. In each of the Resident Tilts a Furrier & his two assistants are to be stationed, and they are to be furnished with twelve Months' Provisions, but eight Months will be sufficient for the Temporary ones. So soon as the Paths are compleatly finished the Furriers, each of whom is to be furnished with a light skiff, is to examine his District for Rubbing-places etc., and erect small Tilts to shelter in during the Otter-season. Those Districts will be for No. 2 Tilt, St. Lewis' Bay & Charles Harbour. No. 4 Alexis Bay & Gilberts Inlet, No. 6 & No. 8 St. Michael's Bay. On the first Monday nearest to the 15th of September all the Marten and Head-traps are to be baited, the Furrier at No. 2 going along No. 1 Path to Deer Harbour, where he will remain the night & return home the next Day; one of his youngsters to go along No. 2 Path to No. 3 Tilt, where he will meet the Furrier from No. 4, who will send one of his youngsters to No. 5, where he will meet the Furrier from No. 6 and so on; by which plan there will always be two men together excepting the man left at home & those in Nos. 1 and 9 Tilts. My reasons for giving each Furrier two youngsters are that there will be so much to do for the first Summer that they will be necessary, and as those Paths will kill so much Furs between the 15 of September & the middle of January, that an additional man will be requisite to stay at home and assist in skinning & scraping. So soon as the run of Martens abates, the additional man must be sent to join the Builders' Crew & they will not be attached to the Furriers again until the middle of September following. The Winter traps are to be struck up so soon as the Martens go out of season, and the Otter-catching is to commence on the breaking up of the Rivers, and will end with the Month of June. Each Furrier will be furnished with a memorandum-book, in which he is [pg. 79] carefully to set down every thing that is worth remembering; he must be saving of his Provisions, & never suffer salted meat to be dressed whilst he has any fresh meat; and he must mind to keep every thing in it's place and in good preservation, as he will be made answerable for all neglects; his youngsters being under his command.

On the arrival of the Shallop in Hawk's River, she is to take in what more necessaries are wanting for the Tilts and proceed to them without loss of time, with such a Crew as the Clerk shall appoint.

So soon as the Builders' Crew are settled at their Winter's quarters, they are to begin their work, which is to be – Stop any convenient River or Brook to catch Salmon, if they have not already passed up it. Catch Codfish for Winter's use. Build a three-masted Schooner of 63 feet Keel, 15 ft Beam x 6 ½ feet Hold. 14 Sealing skiffs, 6 light skiffs of [space] feet Keel, 1 ditto of [space] feet Keel. Some Canoes. Saw Plank & Boards of such sorts as will be wanted. Build a large Pit-fall Trap, if there is a convenient shore for it. Cut and square a frame for the Merchant's house. Take off & cure plenty of rinds the next spring, and do whatever else may be judged best. So soon as the vessel returns from Quebec, she is to be unladed and then sent home. If she could be laden with spars or other useful timber that would help to pay her expenses.

Fifteen Months provisions should be provided for the whole of the 24 Men who are to remain in the Country; and the vessel be provided with as much as will be necessary for her Crew, & the other hands whilst they remain on board; on her departure for England she is to have one Month's allowance, but no more.

Head-traps for Foxes

Bed, square; Jaws, square, equal to the Bed. & they must have strong, sharp teeth. Bridge [ink stain] inches square, turn upon axle-trees, and to keep the Jaws toiled by a pivot on each side resting upon another on each Jaw, and all of them a little flattened where they touch, that the trap may not stand too fickle; A loose "u" spring at each end, with a bow at the angle large enough to admit the finger of a man; Shank one inch longer than the Spring, & a hole in the end of it.

[pg. 80] [Section crossed out entitled "To toil the Head-Trap"]

Dimensions of Mr. Barnetts two Ground-parlours

The largeſt is 17 x 17 x 10-2 feet.
The small one is 13-5 x 13-4 x 10-2 feet.

To toil [tail] the Head-trap

Drive two ſtrong ſtakes firmly into the Ground, at eight inches asunder and five feet long (if upon open Ground, where snow is likely to accumulate) with a short fork on the top of each, and a loose ſtick [pg. 81] across the forks. Cross-band a piece of Bait firmly upon the Bridge. Toil the trap and hang it at a proper height from the Ground (according to the animal you want to catch) by a piece of line tied to the Shank of the Trap & the cross-ſtick of the Gallows. Scarce any Beaſt of prey will be so shy as not to seize the Bait, and that inſtant he will certainly be caught by the head & pull the trap down; the teeth of the Trap will prevent his pulling it off with his feet, and he will not be able to carry it far. No weather can put that trap out of order, and it is toiled much quicker than those which catch by the legs. The beſt Baits for the above Trap are a piece of Salmon; Cod-fish; Deer's paunch; Seal's fat, or any kind of flesh. When that trap is toiled upon open Ground, the Stakes may be touched with Scenting. Half a Seal's carcass, fixed upon a tall pole, is excellent, as that will be apt to attraɕt Ravens, and they will bring Bears, Wolves & Foxes.

A Sailing Sled

Build it in the same manner as the Sleds used by the Esquimeau, Indians. The sides muſt be of two inch Plank, shod with iron; one foot & an half high, 40 feet long, and fixed fifteen feet asunder. It muſt have two maſts, with Schooner sails, & a Boltsprit. Stancheons muſt be fixed upon the sides, to support the maſt-thwarts, which muſt be 1½ inch thick, & the upper-works of ½ inch boards nailed to the Stancheons: For the rudder, see page 103. [pg. 82]

To preserve Potatoes

Pack them in a cask immediately as they are dug up, clearing them of what earth adheres to them, & stowing them in the cask with dry earth or sand below, between and above them.

To preserve Carrots & Turnips

They are to be dug up when properly grown, their tops & tails cut off; freed from what earth adheres to them & packed in casks in the same manner as potatoes. The carrots will require near double the time to boil them, than fresh-drawn ones do.

To carry Boats on board of a Hunting-Schooner

Build them of two sizes, that one may be stowed within the other. By so doing, you may carry double the number which you otherwise could do.

Pitfall-Trap

Those which are intended to catch small animals only, must have a covering-board fixed at a proper height over the falling-doors, that the larger ones may not go on them.

A temporary yard for young Geese

Drive four stakes into the ground, in a square, and run a net round them.

The mode of catching Beavers in a brook, after it is frozen, as practiced by the Indians of Hudson's Bay: communicated by Mr. Paulson, who resided [blank space] years at the Company's settlement on Slude river[16]

Taking care to keep at such a distance from the house as not to disturb the beavers, they cut as many stout [pg. 83] stakes, of dry wood, as they suppose will be sufficient for their purpose, and lay them down at the places where they are to be used; that always is at some thirty yards, or more above the house, for the beavers will not then run down the ſtream, as they cannot then get over their Dam, which is always at no great diſtance below, & they can have the whole range of the brook upwards. If the Day is not too far spent, they go to work immediately; otherwise they return the next morning after the sun is up, as they know that the beavers will then be all at home. The firſt thing is to cut a hole, quite across the brook, with their ice chizzels, & run the ſtakes through it, close to each other, & shove their points firm into the mud at the bottom. They then advance nearer to the house and do the same there, excepting, that they have an opening in the middle, in which they place a net; which is attended by a woman who plucks it up so soon as she feels a beaver ſtrike into it, tosses it upon the ice and immediately kills it. All that business is done with the leaſt noise possible, but all being now ready, they break open the house. On the firſt ſtroke being given, the beavers fly out of it, and one is soon caught in the net, which is quickly replaced. When no more come to the net, if they think that they have not killed the whole family, they cut through the ice all along both shores & try if there are any hollows under the banks, and kill what they find either with spears, or catch them by their tails & toss them out. If what they have killed are in pairs, they suppose that none have escaped, but if they have an odd one, they will not readily give up their search until they have found its mate. Beavers generally bring forth two at a birth, and in general those are male & female, which pair together when they arrive at a proper age: but should both of [pg. 84] them prove of the same sex, they muſt then wander about until they can match themselves and are called hermits so long as they live single. Beavers always bring forth their young in the Spring, before the ice breaks up, and the firſt-born pair generally file off and build a house for themselves, some time about the third July or Auguſt afterwards; but whether they do that through inſtinct, or are driven out by their parents is a matter not easily determined. Be that as it may, the firſt winter the family will consiſt of no more than one pair which are then called Beaver; the next Spring they will have bred, & then they are diſtinguished by the names of a pair of Beaver, & a pair of Pappuses; during the second winter they are called Great Beaver, and small medlers; they will have another pair of Pappuses the second spring, which will be called small medlers in the winter following, and the others Great medlers; the third Spring they will have another pair, which will make the whole Crew, or Family consiſt of eight but in July or August following the first pair of young-ones will quit the house, build one for themselves & become Beaver. Should that not happen, as I have been assured has sometimes been the case, and that as many of sixteen have been found in the same house (which, if that was true, I am much inclined was two houses made into one) they will then increase most rapidly as there will (be) four born the fourth Spring, & six the fifth, which will then make eighteen in the whole; a number much too great for any house; which I have yet seen, to contain. I have heard of their having five, and even [pg. 85] seven at a birth, but it must be observed that very few men of education or enlightened minds ever saw an inhabited beaver-house, and fewer still ever practiced catching them; consequently all these wonderful stories are told by ignorant, illiteral men, and that the majority of those people are extremely addicted to lying.

Although the above indian method answers well yet I am of opinion that the following will be both a surer and more expeditious way.

The best length for Beaver-nets is eight yards; their depth ten or twelve feet; the twine Trall or Seal, and the mesh wide enough to admit the head of the larger beaver with ease. They will be sure to be entangled in such a net and, although they will cut it, they cannot get at the collar-mesh & therefore cannot get out, and as it is well known that a Seal will be drowned in a net nearly as soon as a Dog, it will not be long before that will be the case with them also. Provide four such nets, & make them into two, by seaming two together at one end. Be provided with every other requisite also, & repair to the brook after sun-rise. At the distance of about thirty yards above the house, cut a hole about two feet square through the ice in the centre of the brook; lay the two ends of one pair of the nets, where they are seamed together at that hole, & stretch the nets downwards, one on each side, until their other ends touch the respective banks; there cut other holes of two feet square; half way between each of them and the top-hole cut a hole of one foot diameter and [pg. 86] also four small ones; one on each side about four feet below the top-hole & the others at the same distance below the two round holes, & introduce a stake into each to stand firm. Having overhaled the nets close above the top-hole, lie one end of a line of ten or twelve yards in length to the clue of the head-rope of one of the nets and take a hitch of it round the clue of the foot-rope also, then tie the other end to the thick-end of a pole of thirteen feet in length, & do the same by the other net: the small end of one of those poles is then to be introduced into the top hole and shoved along towards the round hole, & the stake will prevent the current from carrying it too low; should the small end of it appear under the round hole it must be moved to the right or left by the assistance of a forked stick applied to the thick end until it does, when your assistant is to hold it there with a sharp hooked stick; you are then to go to that hole & shove it forward by the help of your forked stick, until the small end appears under the lower large hole, where it is to be pulled out & laid upon the ice. Having done the same by the other net, draw up all the four stakes & proceed to set the net, which will quickly be done by one man haling upon the line & the other putting the net gradually into the water through the upper hole. Having put both nets in run a stick under the head-ropes at their junction & lay it across the hole, then hale each net tought, hook their head-ropes up, through the round holes & put a stick under them across those holes, to prevent the weight of the beavers from sinking them & thereby enabling others to pass over. You are then to put out the other pair of nets in [pg. 87] the same manner halfway between the former and the beaver-house. All that must be done with the least noise possible; & be particularly careful not to speak a word. One man is then to take hold of the head of the lower net with a finger & thumb, to feel when the beavers strike in, and a third man, if you have one, is to do the same by the upper net, & the other is then to break open the house. Upon the first stroke the beavers will fly out & be immediately perceived to strike into the net, and none shall in the course of a few minutes be seen in the angle of the house, you may suppose them all to be caught in the nets; then withdraw the sticks from under the head-ropes at the round-holes, cast off the lower ends, and hale the nets up through the top-hole. Kill such as are still alive, disentangle them, & examine the nets, to see if any have cut through & escaped; but if you find even pairs & none has been felt in the upper nets, you may take that up also; otherwise put the lower-nets again & hunt the banks with your dogs to find where any have concealed themselves & cut through the ice at those places. If you attempt them in October, before the brook is frozen over, be provided with three pairs of nets; two pairs of which are to be fixed above, & the other pair below the house, & the latter set with the angle down the stream. Those nets will be very easily & quickly put out, by means of lines; one end of which must be tossed across to your assistant and they may be fixed close

to the house without the leaſt noise. In taking up the upper nets, by caſt-
ing off their lower ends, the ſtream will draw them both together, but one
end of the lower net muſt be haled over until it joins the other before you
attempt to take it out. [pg. 88]

**Plan of three pairs of nets set in a brook of four yards in width. The
uppermost is supposed to be set under the ice, but the others in open
water, to show both ways. Scale 45 parts of an inch to a yard**

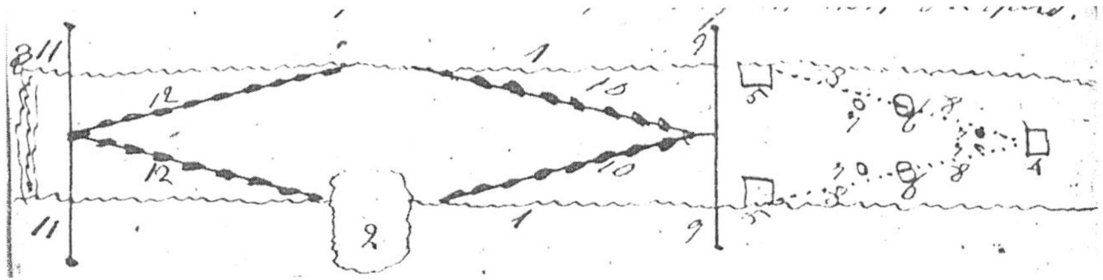

Explanation

1.1. The banks of the Brook. 2. Beaver-house. 3. The Dam, or Stint. 4.
The upper hole through the ice. 5.5. The two lower holes. 6.6. The two
round holes. 7.7.7.7. The four small holes, for the ſtakes to direct the
poles. 8.8.8.8. The dotted lines represent the nets under the ice. 9.9. A
line ſtretched across the brook to support a pair of nets set in open water.
10.10. The two nets, well corked. 11.11. A line ſtretched across the brook,
to keep the angle of the lower-net in it's place. 12.12. The lower nets well
corked.

Extracts from MacKenzie's Voyages[17]

North-Weſt Company eſtablished in the Winter of 1783. The concern was
divided into twenty shares; a proportion of which were held by those who
resided in, and managed the business in Canada, & the reſt by those who
lived far up the Country, to trade with the Indians. The latter contributed neither capital nor credit. Each of those were allowed two shares &
were allowed to retire with one share & sell the other to any young man
in the Company's service, who was approved of. In 1798 the shares were
increased to forty six.

List of Goods for the Indian Trade

Coarse woollen cloths of different kinds; Milled blankets of different sizes; Arms, & Ammunition; twists & cans of Tobaco; common Hardware; Cutlery & Ironmongery of different descriptions; Manchester goods; Linens, & coarse Sheeting; Thread [pg. 89] Lines & Twine, Kettles of brass & copper, & sheet Iron; Silk, & Cotton handkerchiefs; Hats; Shoes & Hose; Callicoes, & printed Cottons, etc. etc. etc.

Skins etc. procured in the course of one of the Years

106,000 Beavers, 2100 Bears, 1500 Foxes, 4000 Kit Foxes, 4600 Otters, 17,000 Musquash, 32,000 Martens, 1800 Mink, 6000 Lynx, 600 Wolverine, 1650 Fisher, 100 Racoons, 3800 Wolves, 700 Elks, 750 Deer, 1200 Deer half dressed, 500 Buffalo robes, & a quantity of Castorum.

Number of Men employed by the Company

Fifty Clerks; seventy one interpreters & Clerks; 1120 Canoe-men, & 35 Guides. Of those 5 Clerks, 18 Guides, & 350 Canoe-men are employed in carrying Goods up the river from 1st May to end of September; for which trip the Guides have from £33-6-8 to £41-13-4; Foreman and Steersman from £16-13-4 to £25-4-. Middle men from £17-4-4 to £14-11-8 besides one Blanket, one Shirt & one pair of Trowsers, together with the liberty of Trade; by which they sometimes double their wages. They are also found in Provisions. Those who wintered up the Country had above double those wages and equipments. Those who were both Clerks & Interpreters had wages from £41-13-4 to £100-5-4 & double equipments. Apprentices were engaged for from 5 to 7 years & were allowed £100 with provisions & clothing. Clerks when out of their apprenticeship, if not otherwise provided for, had from £100 to £300 pr. anm. wages, together with all the necessaries. A large Canoe costs about £12-10-, & will carry 8 or 10 men with 60 paccages of goods, 6 Cwt. Bread, 2 Cwt. Pork, 3 Mash.Pease, etc. The Paccages weigh 90 lbs each, & some men have carried two the distance of 10 miles in six hours. They have a vessel upon each of the three great Lakes, viz. Erie, Huron, & Superior of from 50 to 70 Tons burthen.

An improved way of making Hawk-fences, for entrances into Deer-pounds

Make them of No. 1 iron-wire, twelve feet long and six inches asunder. The outer-ends pointed sharp. They must stand only four feet high, and be concealed by pricking young, well-feathered trees outside of them. The fence of the Pound must be produced opposite to their points, & at the distance of two feet from each Hawk, & a tree with its branches on nailed from Post to Post two feet above the top of the Hawks, & another two feet above that. The points of the Hawks to be [space] asunder, which will be wide enough for any deer to go through, but not return. [pg. 90]

Another way to catch Otters upon a path which runs across a soft Marsh, and which will prove very beneficial, if executed with great care

How many young ones an otter usually has at a time is what I do not yet know, but believe that the number seldom exceed three. At the latter end of the summer several broods of them will collect together, and continue in company for some time; insomuch, that I have seen nineteen so associated: and, as they follow each other, when travelling along a path, it is necessary, that this Pound should be of considerable length, in order to catch the whole of them. Those paths always run from one water to another. Begin the Pound as near to one of those waters as the nature of the ground will admit of. Having determined upon the length of it, drive two stout stakes into the ground as deep as you can, at one end of the Path; each being three feet distant from it: do the same at the other end also, and then run a line from one to the other, parrellel [*sic*] with the Path; but if it runs in a crooked direction, other stakes must be driven in at the several turns, to keep the fence at the distance of three feet from the Path in every part. Other stakes must then be driven in for the whole length on each side of the Path, at no more than two inches distance from each other. Saw all their tops off at four feet above the ground, and nail a batten along them from end to end. The work ought to be commenced about the first of July, and carried on along both sides of the Path — and equally every day, until it is completed; taking particular care [pg. 91] not to leave

any chips etc. in site, nor to tread within the two fences, and rake the bare
parts of the Path with an iron rake, to discover whether or not the otters
venture along it. Having done as much as is already ordered, let it remain
in that ſtate for a month, if you can spare so much time, then lay two
bridge-boards across the centre of the Path, and take care that they will
not move when an otter goes over them. Then nail a ſtrong batten across
each end of the Path, from the top of one end-ſtake to the other and hang
your Door upon that. At a proper diſtance within the Pound, nail another
batten, to which a wooden hook is to be suspended to support the bottom
of the Door in a horizontal position, and hang it up. At the same time
begin to contraƈt the entrance, by driving two ſtakes on each side, next
to the end of each fence. That the Otters may be deceived as much as possible, leſt they take fright and refuse to enter the Path, make the Doors
of ſtrong battens, placed in a perpendicular direƈtion, and face them
with fine boughs of Spruce. By degrees contraƈt each entrance until the
Doors will cover the open parts, & drive ſtakes, close together, into the
ground all along the door-way, & cover their top ends with Earth: which
is absolutely necessary to prevent their scratching a hole under the Door.
All the ſtakes being driven so deep into the Ground that their points
meet with either rock, ice, or water; they cannot burrow under them, and
the tops of them being four feet high, they cannot get over them. You are
then to make Hawk fences from each entrance for a few yards diſtance, by
tucking down the tops of young trees, close together (That ought to have
been done so soon as the side-fences were made, to hide the entrances
and conduƈt the otters into them). Let the Pound remain as [pg. 92] it now
is until you discover that the Otters pass freely through; but if you find
that they avoid the Pound, extend your Hawk fences until you conduƈt
them through. That being clearly ascertained and the catching-time
arrived; divide the Pound into two equal halfs, by driving a row of ſtakes
across the centre, fix the two Bridges (which muſt be well rubbed over
with earth in the firſt inſtance) tie one end of a fine wire to the back of
the hook; which supports the Door, support that wire by laying it over
smoothe ſticks nailed at proper diſtances across the side-fences until
it passes over one direƈt above the Bridge. By the time the firſt Otter is
arrived there, I suppose the laſt will have entered within the Door, and as
he will go upon the Bridge, to try to get forward, his weight will pull the
hook from under the Door, that will fall down, and the whole of them will

be secured. Do the same by the other Door and Bridge and the Pound is
compleated: Before the ice breaks up, in the Spring, the Pound muſt be
examined, to see that every part is in perfeċt order; so soon as you expeċt
the ice will give way, hang up the Doors again, and continue to catch
until you find the skins going out of Season, when the centre ſtakes muſt
be drawn up; the wires taken off; the Doors hung up permanently, and
the Pound left open until the autumnal catching-time arrives. It will be
safeſt to drive a ſtake on the inside of each entrance, that the Doors may
fall within them to prevent the Otters from raising them up with their
noses, for they are wonderfully ſtrong, and will make every effort to get
out of confinement. Where there is a Pond in the Marsh, it will be beſt
to make a Pound on each side of it, to catch the Otters as they go to it, of
course, the Bridges [pg. 93] will be at that end next to the Pond. But if that
Pond is a small one, the easieſt way will be, to ſtake it all round at three
or four feet diſtance from the water, & let them drop in over a projeċting
board placed at the ends of the Path on each side, as direċted in Page 29,
No. 24 of Labrador Companion.
N.B. On examination I find that I have given the same direċtions in Page
1, but these are the cleareſt.

Eskimeau Indians

The Eskimeau are supposed to be descended from the Groenlanders and
have dispersed themselves in Tribes, more or less numerous, all along
the Coaſts of Labrador from Latitude 54° 20' on the Eaſt to that of 60° on
the Weſt side of that Country. They also inhabit the Weſt side of Hudson's-
bay, from the Latitude of 60° to Wager sound, as well as the North side
of Hudson's Straight. As they kill numbers of Whales, Seals and other
marine animals, they procure from them a considerable quantity of
Whale-bone, oil and skins; and their trade would prove very lucrative,
were it conduċted properly. That can never be done were it free to all
adventurers; as the short-sighted policy of those who engaged in it would
soon render it unprofitable and ruin the health and morals of the Indians.
To bring that trade to the degree of perfeċtion which it is capable of it will
be necessary to learn their language and make interpreters; to gain their
confidence; to eſtablish an influence over them; to attend to the preservation of their lives and health; to improve their morals; to teach them to

be industrious, and careful of their oil etc.; to supply them, not only with such goods as they have hitherto required, but also with all others that will add to their comfort and be likely to become absolute necessaries in due time. I shall now proceed to give the best directions, which my experience suggests to me, how all those objects are to be accomplished.

To learn their language and make interpreters, prevail upon a few young Eskimeaux to reside along with you; which will enable you to learn their language, and teach them yours.

To gain their confidence, never appear to suspect them of entertaining hostile intentions towards you; associate with them freely and unarmed; put on your best [pg. 94] looks, and pay proper attention to their women, but more to their children.

To establish an influence over them, observe strict honesty in your dealings; sell them no bad or damaged goods; should any break through the fault of the maker, exchange them for others; never tell them a lie; never be guilty of the least deceit; never give way to passion but observe a constant steadiness in your behaviour; cooly reprimand those who give you offence, and appeal to the rest on the impropriety of his conduct; make them clearly to understand that, let them do what they may, you will never attempt to take the life of one, but will content yourself with refusing to trade with, or holding any communication with them in future; should any commit a fault, confess it and solicit pardon, grant it; and admit him to favour again, but accept no present, as the price of reconciliation; do them every service in your power and administer what relief you can to such as are ill or have received any wound; so soon as you can discover the leading men, attach them to your interest, by associating freely with them and admitting them to your table at times; use your utmost endeavours to please them, but at the same time keep up your consequence, without showing either pride or passion, and you may depend upon succeeding to the utmost of your wishes. I might have comprised all that I have written upon this subject by quoting the few words of our Saviour who commands us "To do unto others, as we wish they should do unto us," for they are applicable to all people, times and circumstances.

As many of them have lost their lives from the hostile attacks of the Mountaineer Indians; some from famine in the Winter, and others from severe colds, too much care cannot be taken to prevent those things from happening in future; the loss of every man contributing to decrease the quantity of goods that otherwise would be obtained. When in Labrador I had the extreme pleasure of making Peace between those Mountaineers and Eskimeaux who lived nearest to me, and the guns, with which I had supplied the latter, contributed much to strengthen the arguments which I made use of to the former. They have lived in great harmony ever since, and are likely to continue to do so. Never, since I supplied them with guns, have they experienced the horrors of famine, as they now [pg. 95] kill plenty of Deer etc. when the seasons proved unprosperous for Whales, Seals, etc in the Fall; which they were not able to do with their Bows, in sufficient numbers. The guns have also made them better walkers and more active. By being driven out of their Winter-houses, so soon as the Spring thaws set in, and compelled to live in their tents until the commencement of the next winter-frosts, they are liable to catch severe colds, which may be greatly prevented, by building them comfortable houses, in which they may reside in all weathers. Those houses will also have another very good effect, as they will require so large a quantity of Oil, to pay for them, as must stimulate them to industry. As the three southernmost tribes were compleatly swept away in the Winter of 1773, by the small-pox, communicated to them by a woman whom I had brought to England in company with another woman, two men and a child who all died of that disorder, the greatest care should be taken, that no person be permitted to embark in a ship going to that Country, who has recently had that disorder; as the infection is retained and communicated in a most wonderful manner. The same caution should also be observed, with respect to those infected with the venereal disease; as that would prove equally destructive to them, although it's operation would be much slower.

Respecting their morals, be sure never give them any spirituous liquors, and it fortunately happens that they have a strong antipathy to them; inculcate strongly the virtues of honesty and veracity; and I must do them the justice to say, that I always found both very prevalent in those uncultivated people, as well as the amiable one of hospitality. In fact, I always found their true character to be, the very reverse of what general report had represented.

To teach them to be industrious and careful to preserve their oil etc., furnish them with such goods as will be conducive to their comfort and are expensive; otherwise, they will not exert themselves [pg. 96] more than they usually have done, or you will not be able to purchase all they have to spare, with the trifling, low-priced articles that have hitherto been sold to them. Trade always creates luxury; luxury, wants; and wants create industry. In addition to the articles with which they have hitherto been supplied build houses, on the plan described in Page 242 of my Labrador Companion; Vats to keep their oil in; Shallops and Sailing-sleds; supply them with Sails; Cordage; Anchors; certain Carpenter's and Smith's tools; good Guns; the best Powder & Shot; Balls; Swan-shot; flints; warm woollen Cloth and every other article which will be of real use and comfort to them. It is a very mistaken notion that the most profit is made upon the worst articles, for more percents can be cleared upon good, than bad ones, and ten tons of oil may be purchased in the same space of time as a single bag full. In all my dealings with them I found, that those articles which cost me the most money, produced the greatest profit, and gave the most satisfaction to them; and that all those which only caught their eye but did not prove of real use, were soon despised.

They know nothing of the Eskimeau character who assert that they have great stupidity and deficiency of intellect, for they have strong natural sense and are quick at learning any thing; they possess great invention, and execute every thing neatly, as clearly appears by all they make; no part of the world producing things better contrived for their several purposes, or better made. I have examined them all with the greatest attention, and was scarce ever able to discover a fault. They are a most amiable, ingenious, tractable and well-disposed race of mortals, and would improve greatly and rapidly by proper management and cultivation.

Goods for the Eskimeau Trade

Houses; Vats for Oil; Hogsheads for ditto; Storehouses for ditto; Shallops; Sails; Cordage; Blocks; Anchors; Butcher's Bone-saws; Pit saws for bones, with a frame; Drills for bones; Hasps & Files for ditto; Guns with barrel 2 ft 8 in: 4½ lb weight, bore balls to the pound, pointed barrels, Patent locks & price from 5 to 5½ Guineas; Canaster[18] Powder; Patent shot; Turn-screws, vices etc. for Guns; Flints; Woollen-cloth of warm quality; Milled Stockings; Barming-glasses;[19] Brimstone; Whaling-guns; Harpoons for Guns; Combs, small & large; Hair-brushes; Salmon-rods with reels/lines, flies, bottom-links, gaff-hooks, etc. Trout rods with all necessaries; Head-traps; Bear-traps, Pick-axes; Spades; Balls for guns; certain Carpenter's, Smith's & Cooper's tools; Powder-flasks; Shot-balls; Water-pails; Oil-cocks; Tin-funnels; Fox & Otter-boards; Jackets of 3 sizes for Men & Women; broad bindings of various colors; Threads; [pg. 97] Oak boxes, with locks, for clothes etc; Sail-needles, Palms & Twine; Planks, Boards;

N.B. 1500 feet of Plank, & 1500 feet of ½ inch board will build a Vat fifty feet long, ten feet wide, & eight feet high, which will hold ninety five Tuns of Oil. All those goods in Page 291 of Lab. Comp.; Sheet-nets for [Grouse?] 45 x 12 feet; mesh 3 in sq.; twine Herring/Duck-nets 120 x 5 feet, mesh 4 in sq. twine, Herring & backed; Seal-nets, & twine to mend with. N.B. The Gun should be in a Box, for which see Page 107. Patent Swan-drops. Kaleidoscope of 5.

Better directions for building a House of boards, than any yet given

Lay Sleepers to support the outer walls and inside Partitions, and as many Beams as will be sufficient for the floors. The Stancheons for the outer Walls must be 6 x 3 inches, and for the Partitions 4 x 2 inches. They must be morticed into the Sleepers, the tops flat, and Wall-plates of Plank or Board tacked upon them, to keep them in their places until the boards are nailed on. All the Boards for the outside of the House must be of the same width, that the Stancheons may be fixed at equal distances. The whole of the Boards for both the out and inside of the House must be planed on one side and their edges neatly jointed. Those on the outside of the House must be nailed perpendicular, and their edges meet on the centre of each Stancheon, but all those in the inside are to be nailed horizontally, by which the Partitions will not require so many Stancheons. All the outside boards being nailed on, and also those on one side of each Partition, begin with the lining-boards; having fixed the bottom one fill the vacant space between the Stancheons with sand, Peat-earth or Moss, then nail on another board, and proceed in the same manner until you come to the top board, when the wall-plate must be taken off, that the vacant spaces may be compleatly filled up, when it is to be nailed firm on. The front wall of the House must be higher than that of the rear, in the proportion of one inch to every ten feet of the breadth, which will be a sufficient fall to carry off the wet, and that will easily be done by laying on additional Wall-plates, properly sloped away. The Rafters are to be of Plank 8 x 2 inches set upon their edges from front to rear; inch Boards are to be nailed length-way upon them, with their edges meeting upon the centre of each; those boards are to be covered with a coat of Sheathing-paper, laid on with hot Pitch & Tar, and a coat of sods over all. A Board muſt be nailed along the outer ends of the Rafters, the vacant spaces between the top of the Wall-plate & the Rafters filled with dry Moss, & lining-boards nailed between each pair of Rafters even with lining-boards beneath. The Door-cases and Window-frames muſt be fixed in their places before the Boards are nailed on, and all seams likely to admit air muſt be covered with a slip of paper paſted over them. For the Fireplaces, Stoves etc. see the directions already given. N.B. If the Stancheons are shouldered at each end & nailed to the Sleepers & Wall-plates, that will save much time. [pg. 98]

To build a Store-house which shall not admit drifting Snow

Lay a frame of Sleepers & mortice into them Stancheons 8 x 3 inches, tack on Wall-plates, to keep them ſteady. They are to ſtand four feet asunder & have half-inch battens of three inches in breadth nail[ed] across them, on the outsides, at the breadth of a board from each other. Joint the edges of the Boards & nail them upon the Battens, & as each is nailed on, drive 1½ inch nails through the Battens into the Board between the Stancheons, which will keep them close together & prevent the drift from penetrating. The Roof is to be conſtructed as direʃted in the laſt Page; the Door-case muſt be lined with Swanskin & the Door edged with it also, & the Key-hole covered. Build a Porch before the Door, of 10 feet in length, in the same manner; but there need not be a Door to that. The Store-door is to open outwards, leſt any drift should get in and prevent it's opening. The Battens may be nailed to the Stancheons with one nail of 1½ inch long at each end only, as the boards muſt be nailed on with 3 inch nails. N.B. If the Stancheons are shouldered at both ends & nailed to the Sleepers & Wall-plates it will save much time.

To discover the exact depth of Water that there is at the bottom of a Cask, or Vat filled with Oil

Provide a glass tube long enough to reach the bottom of the vessel, with the hollow part two tenths of an inch in diameter, & both ends open;

introduce that gradually down to the bottom of the vessel, then ſtop the top end with your Thumb and draw it out, and it will show the depth of water that is under the Oil.

To extract what water there may be at the bottom of a Cask of Oil, without spilling the Cask

Provide a pewter Clyſter-syringe, with as many additional lengths to pipe, to screw on, as will reach from the Bung to the bottom of the Cask; introduce that at the Bung-hole, down to the bottom, & it will draw up nothing but water, so long as there is any in the Cask.

Proportion of Oil to Whale-bone, in Whales

Our Whalers calculate that each Whale will produce one Tun of Oil for every hundred weight of Whale-bone.
Price of Whale-oil in 1717 from 40 to 50 Pounds per tun, and of bone from 76 to 78 Pounds. [pg. 99]

Mountaineer-Indian way of rendering all kinds of animal fat into tallow or oil, in such a manner that it shall not turn rancid, or stink with keeping

They cut off the fat without leaving any flesh upon it, cut it into small pieces, put it into a Copper, or other large vessel, with a [due] proportion of Water, and boil it over a gentle fire, sufficient to keep it boiling slowly, until all the Water is evaporated; which is known, by it's ceasing to bubble, & keep skinning the filth off the top as it rises: They then draw off the clear Tallow or oil, and press the dregs. The English boil their Whale-fat in very large Coppers, with some water at the bottom, for twenty four hours, & then let it ſtand as long to cool. The sooner the fat is rendered, after the animal is killed, the better.

The Mountaineer way of skinning an Otter

That work is always performed by a woman, who is provided with a sort of chizzel, made out of the long bone in the fore-leg of a White bear, with

the corners rounded off. She begins at the Mouth, and works with a sharp
knife until she has got a sufficiency of the skin turned over the head; she
then hangs it upon a ſtrong ſtake or tree, fixed in a perpendicular direction, and then drives the skin off by ſtriking the bone-chizzel very forcibly between the skin & the fat. By that process, she will compleat her work
in about half an hour, & the skin will have little occasion to be scraped,
when dry; but it generally takes three hours to skin one with a knife, and
requires as long a time to scrape it afterwards.

Head-traps for Seals etc.

Some few attempts have been made to catch Seals in Traps, but they
all failed. Having several times had my Meat-net torn & robbed in the
Winter, that could have been done by Seals or Otters only, and therefore I
am of opinion that those animals may be caught in head-traps, suspended
under the ice by a chain, and baited with jerked, or frozen fresh-fish; but
care muſt be taken that the trap does not take the ground at low-water, if
placed within the Tidesway. Beavers may be caught also, by placing the
trap between the house & their Magazine of provisions [pg. 100] and baiting it with Carrot, Turnip or Cabbage preserved in a cellar. In faſt, every
animal may be caught in a head-trap, if you can procure proper baits.
Traps are not good things to catch Deer in, because they beat themselves
to a jelly, and become scarce eatable.

Wolves and their Cubs to catch

To find the cubs, see Page 202 of Labrador Companion; but to know
whether they are old enough to follow their Dam, take off the green-sward
all round the bait that you lay for them, dig up the earth & brush it fine,
with a bunch of boughs, that you may get the impression of the feet of
them, and if you do so, toil eight head-traps, that you may catch the whole
family at one set. Should you see no tracks of Cubs the firſt time, try a
fresh bait and if none come the second time you may suppose that they are
yet too young to travel, & then drawn upon the scent of the old ones; but
if, upon arriving at the kennel, you find the cubs too nimble to be caught,
leave them, and toil head-traps near the place. Mark the trees as you go

on, that you may find your way back; mark a path also a considerable way beyond your bait, and drag back to it. Every fifty yards or so, cut short a small lateral branch of a young tree & stick a small bit of bait upon it. All the head-traps should be toiled near to each other. Be provided with a bag to bring home the live cubs in.

To build permanent Back-tilts

If permanent Back-tilts are built, at such places as you sometimes have occasion to go, they will be found nearly as good as a house; and particularly so if two of them are built facing each other. Make a frame of 1½ inch Plank for the bottom and two ends; set up a ſtrong Poſt at each front corner & lay a ſtout ridgepole from one to the other; make Rafters of Plank 3 x 1½ inches, & lay them at the diſtance of the breadth of a board asunder; nail inch boards upon them, whose edges shall meet at the centre of each rafter, cover the seam with a ſtrip of brown paper, laid on with Paſte, and a batten of ½ inch board over that; give a coat of Pitch and [pg. 101] Tar over the whole (or omit that, as you are disposed) and a top-covering of sods. At the ends set up ſtancheons 3 x 1½ inches, which muſt be shouldered and nailed at both ends, and inch boards nailed upon them, in the same manner as on the Roof, & papered and battened. One of them will make a comfortable lodging, but two form a compleat Indian Whigwham. The Red Indians of Newfoundland build their Winter-houses square and leave a hole in the centre of the Roof for the exit of the smoke, their fire being in the centre of the house. That is a good way to make it warm, and then neither a fireplace nor chimney are wanting, but I think, that an outer-room, with a door to it, would make the fire burn ſtedier [sic], and that the smoke would ascend better & not incommode the house. A house for Furriers, of that sort, should be about eighteen feet square; the Roof muſt be elevated above the Wall-plates about five feet, but the Roof of the outer-room have a slope of two or three inches only. The outer-room should be partitioned off for a ſtore-room; leaving a passage to the inner-room, and a roomy shed built outside of all: The shed should have a Deathfal & Head-trap in it, as it will frequently be entered, by animals of Prey. About 3x5 inch boards of 10 feet in length, & 230 1½ inch Plank would build it.

The Principles upon which a Sailing-sled must be built

The Eskimeau-sled is the best model to build by. Those are twenty one feet long, and about eighteen inches wide; as they are drawn by dogs they are not liable to be upset. Each side is made of a plank of spruce, about ten inches broad and two thick; the foremost ends of which are sloped away from the bottom to the top for about one foot in length, that they may rise over the inequalities of the snow, and they are shod, underneath, with the white bone of a Whale, but whether that is a rib, or a jaw bone I do not know. They are fixed parrellel [*sic*] to each other, and have broad battens sewed across the upper edges of the sides, which project five or six inches beyond them, to give more room. One of those sleds should be procured for a model to make a sailing-sled by. The greater the length (taking care not to make it so long that it shall be liable to break in the middle in case it should sometimes go over such uneven ice [pg. 102] as would leave it hollow underneath) and the broader the bottom, the less it will sink in the snow, and the greater the width the more pressure of sail it would bear without upsetting; and consequently the swifter it would run. To procure breadth at the bottom, without too much weight, fell very stout trees, saw a plank out of the heart of each from four to eight inches thick, according to the length and breadth that the sled is to have, and rub away on each side until the topsides of them are no more than from two to four inches thick; give each of those planks the proper slope at their foremost ends, and the length of that slope must be in proportion to the thickness and breadth of them. If you cannot find trees of a sufficient length, two or more must be joined firmly to each other by a long scarf, and firmly bolted with iron bolts; one end of which must have a broad head & be let into the bottom of the sled, & the other end must have a screw for a nut, to draw it perfectly close. The bottoms must then be shod with the bones of a Whale, or with iron plates. The Whale bones are pinned on with pins made out of the same bones, which wear as the shoing [*sic*] does, but iron plates must be nailed on; and the heads of the nails made flat and let into the plates, that these may not stand proud and impede the velocity of the sled. The height of the sides must be in proportion to the size of the sled, from one foot to two, and they must be fixed with beams, in proportion to the width of the sled, and the foremost sides of them dubbed away, slope way, left they strike forcibly against edges of ice; and one must be placed under the heel of each mast. The top of the sled must be covered with broad battens, and it is to be sailed with two, three or more schooner sails and a Gib. The Masts must not be too high, as every inch of height increases the Leaver, nor must the sails come down too low, as that would be inconvenient in gibing, but they must have a greater breadth than the sails of boats, & be spread either with a spreet or gaff. If in sailing upon a wind if she gripes too much she will require a larger gibb, and if she carries too slack a helm, that will be for want of more after-sail.

[pg. 103] To make the sails properly I advise an attentive perusal of Captn. Gower's ship Transit, in his Supplement to Practical Seamanship, printed for J. Mawman, Poultry, 1807.

Such a sled, if well executed, I am certain will sail at an incredible rate, answer her helm perfectly well, and work quicker than any ship ever did or can do. She ought to be well tried near home before a journey to any distance is undertaken, lest any part should fail, and that should be done upon the uneven ice on the open sea, as well as upon that on Bays or Rivers. More breadth may be given to the bottom of the sled, if that should be found necessary to keep her from sinking too deep in the snow, by nailing pieces of Plank, of 2, or 1½ inches thick, across the bottoms, to project beyond their sides, before the shoing is put on; for, the less she sinks, the swifter she will go.

Head-trap

A better Head-trap than the one described in Page 79 will be a common Single-spring trap, if proper size, with a loose "v" spring fixed at the end opposite to the Shank, and the Bridge made to turn both ways. As that will be hung to the Gallows by the Shank, that will be above the animal's head, when he is caught, and the weight of it will twist his head round, & most effectually prevent his going ten yards from the place.

To steer a sailing-sled

On each side, and not far from the stern, fix a strong Post with a notch cut in the top, the same as in the head of a ship's hand-pump, to fix a leaver in, which is to be hung in the notch by an iron bolt, & hoop those Post[s] with iron, that they may not split. The rudders must be the Leavers that are fixed in those Posts; the inner ends of which must come so near to the center of the Sled as only to give fair room for a Man to stand between them, and their outer ends project about six or seven feet beyond the sides of the sled; they [pg. 104] must be pointed with iron, have strength in proportion to the size of the Sled, so well ballanced [*sic*] that the inner ends shall be but a small matter heavier than the outer ones, and the Posts which they are hung upon only so high that the outer ends of the Leavers shall not be more than about eight inches above the level of the ice when the Steersman holds the handles of the inner ends with his arms hanging down at his ease. When he wants the sled to turn to the right he is to lift up his right hand until the outer point of the Leaver or Rudder presses upon the ice, & the contrary when he wants it to turn to the left, and, as the Sled will turn very quick, he must be almost constantly catching one side or the other, and by so doing he will steer with great exactness. So soon as the drift-banks of snow get quite firm, which they always do immediately after a clear, calm Day or two in March, the Mountaineer Indians are fond of amusing themselves with sliding down the side of a hill, on which there is a good drift-bank, and they perform it thus; they draw their sled up to the top of the hill & place it at the edge of the declivity, and are provided with a stick to steer with, and which they hold in both hands as if it were a double-headed paddle; they then sit down, upon the hinder part of the sled with their feet forward; then they urge the sled gently forward until it gets upon the declivity, when it goes off with great velocity; as it is much disposed to broach too, they tip upon the snow with one end or other of the stick, as occasion requires, and if they steer well, are at the bottom in an instant, but if they make the least mistake the sled broaches too and the rider is rolled off to the great merriment of the lookers on. In that manner they will go on for hours.

Upon mature consideration I am of opinion that when the white bones of a Whale cannot be procured, sailing-sleds may be shod with sheets of thick Copper, cut so broad as to turn up two inches on each side & nailed near their tops; their edges must be filed even, that they may join well. Iron plates may be done in the same manner, but iron will be apt to rust in the summer & thereby soon decay. The turning up of the sides of the

Copper or Iron must be done with great care, that the shoe may be of an equal breadth from end to end, as inequalities will impead [*sic*] the velocity. A small Pit-saw, for bone, would be very useful, and also a Butcher's saw, for cutting it across. [pg. 105]

Salmon oil to Extract

Salmon oil being a most excellent scenting for Traps, it will not be improper to give directions for extracting a sufficient quantity of it for each winter's use. Build a rendering vat of 2 feet square at the top, one foot square at the bottom and about two feet high; place it over a Receiving-vat of 2 feet square & one foot high; make the bottom of the former of inch board & support it on four legs (like a stool) upon the centre of the Receiver & line it with good wrappering [*sic*], which must be made rather larger than the inside of the rendering vat & fixed upon a loose frame which must be placed upon the top frame of the vat, that it may be taken out at pleasure to be emptied. As a great deal of the fat of Salmon will be found upon their gut, that must be carefully picked off, and put into the vat, together with some of the Belly-part of the fish. Put water into the Receiver & draw off the oil in the usual way. Whilst one Vat-ful is rendering, the fat & belly parts of other fish may be thrown into a cask to fill the vat again. Each Furrier must carry a Phyal of that in his pocket to scent his baits with. That and Treacle will be found the two best things that can be used. Or a large bag, the shape of a Jelly bag, may be fixed upon hoop & suspended over a keg with a spigott & faucet to let off the water, and another to draw off the oil.

Pitfal-Traps

On reconsidering Pitfal-Traps I find that it will be so extremely difficult to build a Bolting-yard to them on places exposed to drift, that it will be best not to build one on any place that is likely to be frequented by white-bears, and to build these only upon exposed situations, which are of sufficient size to catch Wolves; the dimensions of which will be Eight feet long, six high and five wide. The Wolves may be killed in the traps by a spear and the Foxes by a bludgeon. Build the sides with 1½ inch Plank

nailed perpendicular on the inside of the Stancheons, and in the manner
directed for Houses on Page 97, that they [may] keep out the drift. Build
a bridge up to the top on each side, & let the ends drift up, if they will.
Drop bits of bait upon both the bridges, to decoy the animals up to the top.
Always run a drag as you walk your round. The best Drags are a piece of
a Deer's Paunch, seal's fat or Salmon, and at every half mile drop a bit of
bait, the size of a small Walnut. Rub the top of the falling-doors well with
your drag, & bloody them when you can. Drop Treacle upon them also,
and smear on a composition of Cheshire-cheese, Fat & Treacle, that they
may stick fast. A friend of mine assured me, that he once saw three Foxes,
some [Foulmarts], Stoats, Hares, Rabbits & Pheasants taken out of one
of these traps at the same time, & all were alive; the carnivorous animals
being too much cowed to kill the others. [pg. 106]

Head-traps

Those for White-bears must be 15 inches square; made strong; teeth one
inch long, & be toiled with the lower side of the Bridge 3 feet 3 inches
from the Ground. Those for Black-bears are to be of the same sizes but
toiled only 2 feet six inches high. Those for wolves & foxes to be one
foot square & teeth ¾ inch & toiled eighteen inches high. The same bait
will do for all of them, but those for White-bears are to be scented with
salmon-oil, & the rest have Treacle smeared over the bait. Shanks of all
one inch longer than the springs, and all to have two springs, & twisting
Bridges.

Bullet-moulds

It will not be good Policy to sell Bullet-moulds to the Indians, as that
might induce them to steal Lead, but the best way will be, to have long
ones made, that will cast several on each side, and a pair of small Shears
to cut them off with. The Guns should be of two sizes only; one of a bore
that will carry seventy balls to the Pound, and a few large, long ones, for
Geese etc, of a large bore. All Balls should have tin in them to prevent
their being flattened on striking a bone; the proportions of which should
be one pound of tin to eight pounds of Lead. During the bad Days in the
Winter, a sufficient number of Balls may be cast for the next Summer's

trade. In making the Moulds, the greatest care must be taken that the Balls be perfectly round, and none of them too large for the Guns.

The prices at which Eskimeau Goods should be rated, and English Goods at prime cost given in Exchange for them

Oil 10/per Hogshead. Prime silver Foxes five shill.; inferior ones, two for one; Crosses, 3 for one; Yellows, 4 for one; Whites & Blues, 6 for one; Whale-bone, 4 per Cwt.; large seal-skins ditto; young ditto; otters 1/. White Bears 1/6; Black-bears /6. Wolves 1/. N.B. They never kill Black-Bears in season. Eighteen pence, or two shillings must be the price of a large White-Bear in full season, but smaller ones, or shagged ones must be rated lower in proportion.

Observations on Houses built with Planks or Boards

Although I have given full directions for building Houses with Planks or Boards, yet I have not given my reasons for building them differently, and that I will now do. If the outside of the Walls have the Planks nailed on perpendicularly, with their edges meeting upon the centre of the Stancheons, the Materials, with which the Wall is lined, cannot run out through the seams; but as that way will require full three times the number of Stancheons as would be requisite were the Planks nailed on horizontally, and in that case the materials [pg. 107] of the lining might run out, in case the Planks shrunk much, it would be necessary to provide strips of strong, brown paper, of two inches in breadth, and paste the upper side of them upon the under side of each Plank: by doing that, the lower side of each slip of paper would hang over the seam below it; the stuffing (which should be either Sand, fine Gravel, dry Peat-earth, Sawdust, Moss, the down of Eider-ducks, or any other material of like quality) would press the loose half of the paper down, and it would not crack if the Planks should shrink. The use of that paper is, to prevent fine material from running out through the seams in case of the Planks shrinking, but there will be no occasion for it if Moss or the down of the Eider-duck is used. As the inner Planking must be nailed on horizontally, or the vacancies between the two cannot be filled up or the frost kept out, the slips of paper are absolutely necessary upon them when either Sand, fine Gravel, Peat-earth, or Saw-dust are used. I have given directions for building houses in a great variety of ways, that every Man may avail himself of the materials which he is possessed of, and build in such a manner as the situation requires.

Gun-box, & its contents

The Box to be made of Oak, with hinges & lock. The Gun 2 ft 8 inches long; Scowering-rod of same length, with a worm to screw on; a hand-vice screw to take off the spring; a turn-screw, shape and size as under; a Powder flask of tin, with spring-top; a double-snake for shot; a Bandolier; Punch for Wadding; a Cartridge former; a tin shot-measure; 4 quires of Cap-paper for Cartridges; 100 best Flints of proper size for the Gun; all vacancies crammed full of Fluids to clean with. There should also be an ammunition box of Deal with hinges & locks, which is to contain six Canisters of the best powder; 2 bags of No.1 Patent shot; 1 ditto Patent Swan shot; and one ditto of 100 balls, with one eighth part of tin melted with the Lead. Old Boots, & pieces of refuse seal-skin will do to punch the wads out of.

N.B. A representation of the turn-screw is given on Page 108.[20]

Pit Trap for Wolves & Foxes

This Trap must be ten feet long, eight feet broad and four feet high. Make a Ground and top-frame of 1½ inch Planks; mortice Stancheons into them, & nail on inch Boards on the inside.

[pg. 108] Cover the top with inch Boards, but leave a hole in the centre of it two feet six inches wide. Direct under that hole a Board of half-inch Deal, three feet square, must be placed; nails of one inch in length, and finely pointed with a file, must be driven through it, at one inch square asunder, and then other half-inch boards nailed cross-way of the others & over the heads of the nails. Bloody the upper side of that Board well, and place a few Seals'-carcasses at the ends of the Trap. Any Wolf or Fox that jumps down will receive so much injury from the points of the Nails running into his feet, that he will not tread upon it again, and consequently he cannot jump out again. Upon the top-boards of the Trap spill some Blood, Whale or Seal Oil and a little Treacle. If Stones are to be had, build a wall with them close to the outsides of the Trap and slope away with earth, but make the slope of Earth only, if ſtones are not to be had. The only use of ſtones are, in case a Wolf or Fox should gnaw through the Boards. This Trap is to be built where it will be exposed to the drifting of the snow, but no drift will lodge in it. That a man may go into the Trap without being injured by the points of the nails, which will ſtand half an inch above the Board, a ſtool muſt be made, with legs only one inch long, and placed occasionally upon the armed board within the points of the Nails.

A Deer-leap, for the entrance into a Pound

Build the Pound thirty feet wide, at the leaſt, and not less than seven feet high; but the higher the better. In the centre of each end build a platform ten feet wide and fourteen feet long, projecting into the Pound; one end of which muſt reſt upon the top of the Pound, and the other end upon a ſtrong Beam, fixed upon two Poſts. On the outside of the Pound a bridge muſt be built, the top of which muſt reſt upon a Beam fixed close to the top of the Pound and have a gradual slope to the ground, that the Deer may walk easily up it. Both the Bridge and the Platform muſt be conſtructed of trunks of trees, the thickness of a man's thigh (that they may not bend) laid close together and well covered with sods. The Hawk-fences muſt be built close on each side of the Bridges, and then extend to [pg. 109] a considerable diſtance, and widen all the way. These fences will collect the Deer as they approach the Pound; they will ascend the Bridges; advance upon the Platforms, and leap into the Pound. When they are in the Pound they will not leap out of it upon the Bridges, but will run under, or by the side of them to the end of the pound, and then turn back. The outer end of the Platform, which will reſt upon the tops of the ends of the Pound had beſt be built one foot higher than the inner ends. Mind to build all firm, that the weight of the Deer may not make any part bend or shake. As the sureſt time for catching Deer is in the Fall, when they draw out of the Woods and seek the Barrens, one leap, built at the end next to the Woods, will be sufficient in general. One at the other end may be added, if found wanting. This I take to be a better way than those directed before. All Parks in England are furnished with Deer-leaps, and they are found to answer the end proposed.

This alphabetically ordered list in George Cartwright's handwriting is copied into a narrow 17 by 5 cm notebook. Unknown is whether it dates to the period 1770–86 or was compiled later, but Cartwright would have used such lists while preparing for his seasons in Labrador. Many items correspond with information found in the Journal, *and as an historic list it gives some insight into the material requirements of an eighteenth-century merchant station.*

"A"
Anchors
Awl-blades
Augurs
Adzes
Apple-trees E&A
Anvils
Ale strong
Anchovies Essence of
Almonds
Aqua Fortis
Apples, dried A
Axles for Grindstones
Air guns for Ball & Shot
[Alagar]
Antimony, crude
Anise Seeds Oil of
Axes, Broad

"B"
Bread
Brandy
Books for Accounts
Blocks of sorts
Beef
Buttons
Bellows Forge
Bullet-moulds
Brads
Barley Common, Scotch
Brushes Sweeping, Paint
Brad-awls
Boards 1 inch E&A, ½ inch E&A,
    ¼ inch E&A
Blankets
Beads
Brass Pans for Indians

Flies, Turkey
Funnels for Casks, for Bottles
Feathers of all sorts for fishing
    flies
Fenders for ſtoves
Fullers Earth
Fire-irons
Fearnothing (thickeſt kind)

"G"
Guns, Double, L&S Single, air
Gun-powder 3 ſtrong
Gooseberry plants
Garden Seeds E&A
Gimblets
Ginger, Essence of
Gallon Measure
Gauging-rod
Gut, Silkworm
Glew, & pots for
Grindſtones, large & small &
    winches
Glass, for windows
Gum, arabic, guaiacum
Grid-irons, Large & small
Gartering
Gin, Hollands
Gaff-hooks
Gun swivels with locks
Gun, Boxes with implemts.
Gun-caps (Dog's skins)

"H"
Hounds, Htg. & Fox
Hawks, Hoods, Bells
Hammers of sorts
Hatchets of sorts

Hogsheads E&A
Household Furniture
Harpoons for Whales
Hoops E&A
Houses frames of A
Hatchet-helves, of Ash
Hats
Hooks & gimmels
Hinges
Horses
Honey
Hay seeds
Herbs, dried
Hasps & Staples
Harrows
Haws for seed
Hooks, for Fish, of all sorts
Hatcase
Hawsors
Hackles of cocks
Horse sinew
Horse-hair, long and ſtrong [used
    for fishing line]

"I"
Ink, Writing, Indian, Red
Iron, bars of
Inſtruments Surgical
Imps [feathers grafted to repair or
    improve Hawk's flight]
Ising-glass
Iron plate for Sled-bottom, 24 ft x
    2 in., and for other purposes
Ice-moulds
Indigo
Irons, Smoothing
Indian Corn A.

"J"
Jars for Pickles
Jack, roasting (chamber)
Jews-harps
Junk
Jack screw

"K"
Kayane Pepper A
Knives, Splitting, Pen, Furriers
Kettles of Tin
Kitchen Furniture
Knives, Stationers, Butchers,
    Cutting for Hay, Splitting
Kegs
Keys, for saws
Knapsacks

"L"
Lead, pigs of
Leather
Lines for Cod, whitings
Locks for Doors, Chests, [?]
Lemon, concrete acid of, Pure
    Essence of
Lines for Salmon etc.
Ladles, of iron for lead
Lime, quick, E&A
Lamps for oil
Linen, Table
Lanterns of Glass, Tin, Wire
Lines for Curtains, Bell-lines
Lanscape [sic] Painter (Padley)
Lamb-skins, newly dropped
Linen (Frock) for Tents
Liburnum seeds
Lance-bunt

Listing-slippers
Listing [narrow strip of wood]

"M"
Melasses [sic] A
Medicines, Chest of
Mustard
Mushroom powder of
Measures for Gallon, ½ Ditto,
    Quart, Pint
Mutton, Legs of Pickled
Mills, hand for Coffee etc.
Mittens, splitting
Mortars of Iron, Brass, Marble
Mattoc [sic]
Musk Grains of

"N"
Needles Sail, Sewing
Nets, of various sorts
Nails of sorts
Nux vomica
Newfoundland Dogs
Naval Architecture
Nett Duties (Langham's)

"O"
Oil Olive
Ochre Red
Oil Lineseed for Paint & cold-
    drawn
Oil-skin, Silk & Linen
Oats A.
Oven side & Camp
Oziers [a willow used for basketry,
    chair seats, fish weirs, cages,
    etc.]

Oven Doors, Utensils
Oils, Essential for scenting
Ozier Rods

"P"
Pease
Pullies of sorts
Powder flasks
Pepper black, Cayenne
Palms Tail
Pork I
Pigs A
Poultry A
Plank 2 in. & 1 ½ in.
Paper Writing, Strong brown, Cap,
    Window
Pitch A
Pivots & eyes for Traps
Pots Iron for Oil, Cooking
Potatoes in Boxes E&A, Seeds
Pick Axes
Pasteboard
Pease blue or white
Posthorns of tin
Paper for Rooms, bordering
Plough, a light
Pues [sic]
Pipes – tobacco
Pudding-pans of tin
Paints of sorts viz.
White
Red, bright & com.
Blue, sky
Green Light
Ditto Bottle
Yellow bright

Ditto common
Perry [fermented pear juice]
Phyals, for Scentings
Paint-pans of tin
Potting Pots
Pickle Jars
Pails of Tin
Perambulator, fitted upon a Wheel-
    barrow
Pencils, Camel's hair, Black-lead
Primp-berries
Pipkins
Pansions
Pickles of sorts
Parallel ruler, Large & small
Pins of sorts
Plates of Iron 24 ft x 2 in.
Porringers of tin

"Q"
Quart Pots of tin, Measures
Quadrants
Quilting, White Mercelline

"R"
Rum E&A
Resin
Rice A
Reels for Fish-rods E
Rudder-irons
Rugs
Rakes Garden, Hay
Ramrods, spare, Plane for
Rushes for Coopers
Riddles
Rasins [sic]

Rasps
Rings for Curtains etc. both large
    & small, for Bell-lines
Roller, Iron for oil
Rhodium, Oil of [rhodium is a
    metallic element; oil used as
    lure or bait]
Railing, of Cast Iron
Rockets, Sky, Chain
Rag-stones and Charwood [sic]
    Forest ditto [may refer to rough
    building stone]

"S"
Shots Nos. 1.3.5.q.
Spaniels
Shoes
Shirts
Sugar Lump, Raw E&A
Salt Common, Petre, Glauber's,
    Epsom
Shot bags
Salting-pans
Shoes
Spices
Soy
Spades
Spears for Whales etc.
Shovels
Saws, Whip, X Cut single, Hand &
    Hack, Stone, Lock
Scythes
Sickles
Stoves close & open
Slops for servants
Stockings strong & warm

Spoons, metal
Spectacles
Swan-skin
Shot-moulds
Strawberry plants, Seed
Scales & Weights
Sheet yards
Steel in Bars
Snow-shoes A
Spring-Corkscrew for nails
Screws
Scissors
Starch
Screws of sorts
Screw – plated & taps
Sheep A
Still
Soap, hard & soft
Screw-press for Furs
[?] seed
Screw-press for Seals fat
Stag-hounds
[?]
Steels, Butchers
Sparrow-bills
Solder, & Irons for
Sallery [sic], Seed & dried leaves
Spickets & Forcets [may refer to
    spicules and forceps]
Saw Butcher's
Sopha, Cushions for
Silks of sorts, for nets & fishing
    tackle
Salamander [possibly a cooking
    utensil for browning]
Spirits of Wine

Slippers of listing
Seeds of Grapes
Shafts for Carts
Silk worm gut
Spikes & half-spikes
Scrapers for Shoes
Sheep Powder (Twinfen's), Stretton

"T"
Thimbles, Mens, Womens
Turn-screws
Treacle
Tierces E&A
Turpentine Spirits of
Turmaric
Twine Stopper, Shoal-net, Trall,
    Salmon, capelin, Herring very
    fine, Tail
Tin ware
Thread Sewing of sorts
Tripe, Pickled
Tar A
Terriers, true vermine
Ticken for beds
Tobacco
Telescopes pocket
Tape linen of sorts
Trap Hoops for Hogsheads, Tierces,
    Nails for ditto
Tea E&A
Tongues, Pickled E&A
Trowel, Brick-layers
Tin, Sheets of, powder of, Native
Thebaic Tincture
Tassels for Curtains
Turbeth mineral [possibly used as
    a purgative]

Twist, Gold & silver for flies
Time-piece
Tiles, Dutch for Chimney, - pieces
Tenter-hooks
Tallow
Tinder-boxes

"U"
Udders pickled
Umbrellas

"V"
Vinegar of sorts
Vice Rench, smaller
Vitriol Spirits of [sulfuric acid;
    Cartwright suggests it as a
    repellent]
Verjuice [acid juice of unripe fruit
    or crabapples]

"W"
Wire
Wine
Windows
Wheelbarrows, Wheels for,
    Gudgems [gudgeons or pivot
    socket] for
Wafers
Wax, Dutch sealing, Bees yellow
Wick for Candles, for Lamps
Warming-pans
Women's apparel
Wheat A
Worsted
Whetstones of sorts viz.
Charwood Forest
Norway Raj

Grit for scythes
Web, for Chairs & Sopha
Weed, Indian for fishing
Wheels, for Carts
Waggon, A light & Horse

"X Y Z"
Yeast, dried
Yoke for Pails

[Final entry in index notebook]

"Servants"

Boat builder	1
Joiner	1
Coopers	2
Bricklayer	1
Boatsmaster	1
Sealers	2
Wire-worker	1
Smith - white	1
Sawyers	2
Taylor	1
Cook	1
Dairy-maid	1
Tin-man	1
Farmer	1
Laundry Maid	1
Char	1 [possibly a charwoman, to clean]
Furriers	2
Youngsters	2
Gardener	1

Total	30

In this essay Cartwright gives advice to someone who may have had plans to work in Labrador or in the North American fur trade. He instructs on appropriate conduct under various circumstances. In essence a distillation of his own lessons learned, observations, and personal ethics, it is also a brief depiction of a trader's life.

The more useful you make yourself, the sooner you will obtain a Government.

Learn the Eskimeau language.

Learn surveying and the art of laying them down exactly upon paper.

Provide yourself with all the necessary instruments.

Provide a Quadrant & an Ephemeris to take Latitudes & Longitudes, and take Daily observations during the passage.

Learn Navigation, which you may do upon the passage, & be provided with the necessary books.

Keep a regular Diary of all transactions & events. If you describe birds, give the weights of them.

Associate very much with Indians of all Nations, for you will learn something from each. If they displease you talk to them coolly & firmly, but never fly into a passion with them. The more you Excel them in sporting, the right opinions they will have of you. Do not fish for salmon when an Indian is present until you are perfect in the art, lest one should [break] you. For the same reason, you must not fish with a single gut; until you have killed several with flies tied upon three.

Provide a good Rifle, that will carry Point-blank two hundred yards and practise constantly until you are sure of your object. In shooting at Deer etc. always aim at the heart, unless near enough to be sure of the scull.

Provide a long, large Duck-gun (John Bingham of Birmingham makes excellent ones at £3-10- each). Use none but the very best Powder and Patent shot, and put 1/8 of Tin in your balls that they may not be flattened by bones; that is, one pound of Tin to eight pounds of Lead.

In all your conversations with Indians be sure never to tell them that which is not true, nor even such truths as they will not readily believe and you cannot prove. In your dealings with them, never cheat them by imposing bad articles upon them. Should any article break through the fault of the maker (a thing that often happens amongst low priced hardware) exchange it for a new one.

Be very careful not to get into quarrels with your brother officers; they generally are occasioned by drinking too freely, unguarded expressions, too much familiarity, disputing with warmth etc; should any one take liberties with you, keep him at a distance by cool civility without showing contempt.

Should any Man be much inclined to quarrel, the best way to cure him is to send him to Coventry.

Provide a Dog & Bitch from the King's Stag-hounds; no matter how old or disabled as you may get a breed from them. Pointers will be of little use, but spaniels & strong, true-vermine Terriers will be of service, & all must be broke with the check-collar, to drop to the Gun & the word. Always shoot with Cartridges as they cannot lead a Gun & you will load remarkably quiet, which is a material point during the severe frosts. Mould shot is excellent for Deer if you shoot with a plain Gun, but a single ball is to be fired out of a Rifle.

Learn to do every thing with your own hands, that you may know when others do their work well. It is a false idea that a Gentleman disgraces himself by making use of his Limbs, for those things only are disgraceful which are contrary to the Laws of God & Man.

Spread & scrape all the Furs, which you kill, yourself, and write your name close above the tail, that the Company may distinguish yours from others; for you should take all opportunities of showing your excellencies without appearing to do it designedly.

Take accurate surveys of all Bays, Rivers, Harbours etc. which have not been previously surveyed & transmit copies to the Company. When you go to an Eskimeau settlement, count the number of Houses or Tents and ascertain the number of inhabitants, what number of Whales, Seals etc. they usually kill as well as what numbers they might kill were they to exert themselves to the utmoſt: also gain every other information that you can and report those things to the Company. Where you find Tents only, that is a summer, or occasional settlement, for they live in Houses during the Winter. The calculation of the proportion of oil to Whalebone is, one ton of oil to every hundred weight of bone; consequently a large Whale which will produce 20 Cwt. of bone will produce 20 Tuns of oil also. Enquire the time of the year that they kill Whales etc. and whether at their Winter or Summer Settlements. Have a sharp eye to the woods & never fail remarking the size and various kinds of Trees. In short, observe everything, and carefully enter all your observations in your Diary.

Incredible numbers of Eider-Ducks (called Dunter Geese in H.B.) breed upon the small islands & they cover their eggs with their own down; collect that in bags, to bed under the bridges of your traps in the winter. For want of that, use the common short, green Moss. Make a large collection of feathers, for flies, particularly of Peacock's tail-feathers and hackles of all colors and sizes; the large ones being for Salmon. Provide a bunch of the ſtouteſt Gut for Salmon, as well as finer for Trouts.

Provide six pairs of large liſting shoes, have them made broad and square at the toes, and sew a piece of the tickeſt [sic], warmeſt Flannel or warm Cloth to the upper part before and another all round the hind part, that they may defend your inſtep and ancles [sic], and wear a pair of Indian Mockissans [sic] over them. If your feet are not well defended from the froſt, or if they are closely confined in your shoes, your toes will moſt certainly be froſt-burnt. Have a pair of Mockissans made of the skin of a Hind killed in Auguſt, with the hair on the outside, to put on occasionally, when creeping up to get a shot at Deer, Geese, etc. Your dress for Summer shooting should be of dark green cloth, Mohair buttons, & a covering for your hat; the crown of which should be no higher than necessary. White is beſt for Winter, & that may be made more so by putting over all, a clean smock-frock of linen, & a covering of the same for your hat. To keep them clean the longer, carry them in a bag, upon your sled. Be very careful of going to a diſtance from home by yourself, during the firſt winter, and do not rely upon your own opinion of the weather. Never go out of the House without a good hatchet, tinder & matches, as your life may often depend

on the possession of them. Have a side-pocket in your Breeches, and
always carry a sharp Butcher's knife in it. Never let a Deer or Bear get cold
before you paunch it. Skin & quarter them immediately if it be possible,
but that you cannot do during severe frost. A short pocket spy-glass will
be very useful. Abstain from the use of Tobacco, for it is a vile practise and
of no real use. Never drink a dram unless medicinally, and then a very
small quantity will be sufficient. Spirits, although diluted with water, are
not wholesome and weaken the eyes greatly. Provide a three ounce phyal
of Thebaic Tincture[21] & an ounce of Citrine ointment[22] for your eyes, and
Camel-hair pencils to apply them with. In case of being snow-blind, or
catching an opthalmia, foment your eyes with the steam of boiling water
until the severe pain is abated, which will be in three or four Days, during
which time they must be kept well covered, and then apply the Citrine &
Thebaic and do not be persuaded to use any other remedies unless it be a
small quantity of the common Putty-ointment to obtain immediate relief
from excessive pain. When very thirsty with walking, scrape your tongue
with your knife, wash your gums with your finger, gargle your mouth &
throat with water, but swallow only one small mouthful of it. The cause of
thirst is thick phlegm adhering to your tongue & the whole inside of your
mouth from the heat of your breath, which a river of water will not wash
off by drinking it; warm acid liquors will take it off quickly, but the other
is an excellent way.

When faint & weak with walking & long fasting, break a cake of bread
& put it into water, whilst it is soaking, scrape your tongue & wash your
mouth as above directed, then eat the bread & drink one mouthful of
water only. On arriving at home, never indulge in taking a drink or
cold liquor within an hour, but take the chill off it with a hot poker, or a
mixture of what is warm. Never eat much, or one bit of flesh-meat, until
you have finished your Days journey, for it will oppress you and cause
unquenchable thirst. Spruce beer is the best common drink, and an
excellent anti-scorbutic. Be not sparing of either spruce or treacle. Brew
it in vessels that will hold only a week's consumption & have one cask or
keg to succeed another; it will be ripe for drinking in a week. If you begin
to brew where no grounds of a former cask are to be had, put a bottle
of brisk Porter into a Hhd. [hogshead] or part of one into a lesser cask.
Excellent soft Bread is to be made with Leven, but take care that what you

keep for the next baking does not get sour. To prevent that, roll it in flour & wrap it in a clean cloth; defend it from Froſt in the Winter and from too much heat in Summer.

Coxwells' concrete acid of Lemon and Kayan pepper are two excellent things, and if you are fond of them & the H.B.C. send none out, you muſt provide yourself with them. Drawing paper for surveys will be necessary and you will advance your intereſt greatly by making yourself much of that Art. In marking the Soundings in Bays, Harbours etc. do it regularly across, from Point to Point. Gales of wind will discover many small Shoals & Rocks which the lead will miss. Be always very careful how you carry sail in small boats, for they are upset in an inſtant, and therefore the oars are safeſt in squally or ſtrong winds. If it blows so ſtrong that you cannot row ahead, tis time to land, hale up your boat & wait for less wind; as you cannot shew sail to it, but at the peril of your life: and when you can safely turn to windward, you will always row a head faſter. Never let any Man persuade you to do that which your own judgement tells you is wrong; and run no risks, when they are not absolutely necessary. When you get into great Danger, deliberate only on the beſt means of extricating yourself & ſtick to that without changing, unless circumſtances alter. The Man who looks Danger baldly in the face will escape it nine times in ten but the undetermined seldom will. He who wants presence of Mind & resolution in danger shows the ſtrongeſt marks of cowardice, and increases his danger very considerably. Foolhardiness is no proof of courage but a very great one of want of sense.

Long Rifles are bad things, from 2 feet to 2 feet 3 inches are long enough. The heavier the barrel, the farther they will generally send the ball Point Blank as they will bear more Powder than light ones. Rifles which carry a ball of 20 to the Pound will not carry so far as those that have balls of 32 to the Pound, and that sized ball will not permit a Bear or Deer to run far if it is properly placed, & will go through eight or ten Deer ſtanding side by side at the diſtance of 100 yards, if it meets with no bones ſtronger than ribs. Nos. 5 & 9 shot will be very useful, but I do not believe the Company send out any so small. A sliding Powder measure is necessary for a Rifle, and a measure to screw on to the end of the Ramrod, to load the gun with, that no Powder may lodge in the grooves of the Rifle.

When you meet with any of the small, black Brandy-cherries, preserve some of ſtones in dry sand or Bran after rubbing them dry; plant some of them so soon as you land & keep the rest until the Spring. Do the same by good Gooseberries, strawberries or other fruits that will grow from Seed, those kinds must be squeezed in a cloth, to take out most of the moisture, then dried in the air & preserved as above; they are to be sowed along with the Sand or Bran, very thin. If you want to carry out young trees, have them planted in Baskets at the proper time of the year, you may then set those Baskets in Ground where they are to grow, the instant of your arrival & trouble yourself no more about them, for the Baskets will rot & the trees keep growing on.

Never defer that until Tomorrow, which you can do to-day. If you want accurate information upon any points employ yourself to procure it. Constantly keep in mind the admirable injunction of our Blessed Saviour "Do unto others as you would they should do unto you." You will find it will answer best in the long-run and keep you continually upon good terms with your own Conscience; a friend, whom you ought never to quarrel with. Make it your invariable rule to speak civily to every one; for your superiors and equals will not brook rudeness and your inferiors will detest you for it. You will always be better served by those who like, than by those who fear you. Civility is not like money or other worldly valuables, for the more extravagant you are of it, the faster it increases.

Unbroke Spaniels & Terriers will be as good as Beagles to hunt Hares upon small Islands. Let them take their own way & shoot the Hare as they pass.

If you meet with much Field-ice, in the Bay & grapple to it do not go far from the Ship lest the ice parts & prevents your getting on board again. Where you find too much Covert upon any Island burn it so soon as dry after a good deal of Rain. Never set woods on fire after a long dry time, for you will burn all the soil.

*The correspondence in the new Cartwright papers begins with eight letters writ-
ten by George Cartwright in 1771 from Labrador to family members. A 1774 letter
from Daniel Sutton concerns the death of the Inuit boy Noozelliack. Another 1774
letter from John Cartwright, mentioning George, is written to their sister Catherine (Kitty).*

*An 1818 letter from Henry Hunt to an unknown party about transfer of owner-
ship of the Sandwich Bay posts alludes to the profits made by Noble and Pinson in
their dealings. A letter from the Hudson's Bay Company to George Cartwright, also
written in 1818, has a note in Cartwright's handwriting at bottom.*

*Three letters from the mid-1800s concern Frances Dorothy Cartwright's efforts
to have a memorial placed in Cartwright, Labrador, with the help of Bishop Feild,
and one family letter makes reference to George Cartwright.*

*Finally two twentieth-century letters refer to the loan of a Cartwright docu-
ment to Wilfred Grenfell. The "Georgy" and "George Cartwright" mentioned in the
later letters is a Cartwright descendant.*

(Fold lines and edge tearing have obliterated some text.)

Letter written by George Cartwright to Anthony Eyre Esq.

Dear Sir.   Ranger Lodge, 2 June 1771

I am ashamed at not writing to you last year by Mr. Lucas. I'll tell you no
lies about the matter, but frankly confess that the number of letters I had
to write and short time for so doing caused you to be forgotten amongst
many others of my friends for I defer'd setting my hand to paper until my
arrival at Fogo intending to go there with Lucas but I suddenly changed

my resolution the day before he sailed. It was sometime before I could
bring the Esquimaux Indians left with me by Lucas to tollerable behaviour, or get the better of the dread they had that I should murder them for
every slight occasion; but I fancy now they have as great a regard for, and
as little dread of me as their own country people. Tho' at the same time
they are in great awe and carefully avoid doing any thing which they think
will offend me. They were very useful in the winter, for my poor dogs
would have been starved but for the Seal they continually brought up from
our post near the Cape which is ten good miles from hence. They brought
it on a Slide drawn by dogs; and 'tis amazing what they would draw. The
old man who weighs full fourteen stone one day took down eight two-inch
Larch plank about 20 feet long by 11 inches broad and rode on it himself with only three old dogs & four whelps of six months old. The dogs
are about the size of your spaniels. They are a restless sort of people, as
indeed those who live on what they kill must be; being obliged to shift
their habitation with the seasons in order to be near their food. [This
fall they] pitched their tent in a cove near the Cape, but the weather soon
made them think of quitting that: they then applied to me [so] I gave them
nails and lent them a boat with a couple of men to cut & carry wood down
to build themselves an house: which they had no sooner compleated than
the old fellow complained it was so cold 'twould kill his wives and children. One of the latter did die. I then gave them a house in the river about
half a mile below this, where they stayed till the beginning of Febr^y. when
they went down to Lion Head and dug themselves out a house in the snow,
which was the most curious thing I ever saw. Twas exactly of the shape of
an oven with the mouth rather leaning outwards for the conveniency of
entrance, which had a long porch before it. The latter was made of square
pieces of snow cut with a handsaw & water poured over the seams, which
instantly freezing made it perfectly wind and weather proof: the former
was ha[cked out] of the snow with an adze and a square hole cut out over
the door for a wi[ndow] on which was laid a flat, transparent piece of ice,
shaved very thin with [a] knife & cemented with water. The roof also
being not above a foot thick admit[ted] so much light that it was by much
the lightest appartment I ever was in, & had a particular brilliancy from
the numberless small icicles formed by their breath; and without any
fire 'twas warm enough for the tenderest person I know to live in. You
ascended into it by steps which made the entrance easy without stooping

altho' the door was not above two feet square. Somewhat about one half of
the floor was left chair height above the rest & was covered with deer-
skins on which they slept & sat and exactly fronting the door, which was
stopt at night with a piece of ice made to fit exactly to raise the degree of
warmth. On either side the door was a lamp fixed on sticks. The curiosity
& excellent contrivance of these lamps was well worth notice and I would
[describe it for] your satisfaction. It was a large, thick flat stone hollowed
[like a] soup-plate towards one edge, decreasing to nothing on the other,
[paper damage] shelving towards them, the oil all settled in the deep
part, in which [was laid a] small bit of dry moss of a particular sort, plentiful enough on [paper damage] that was their wick: across the lamp some
small matter higher than the wick & a little behind it was placed a small
stick over which hung a number of small rashers of Seal's fat which gently
melting with the heat of the wick dropped down & supplied the consumption of oil. Each wife had her lamp & under them upon the floor lay Seal's
carcasses which they got from our crew & eat either raw or boiled; nor did
a piece which they trampled underfoot for a week come amiss to them:
– They are, to be sure, the nastiest feeders & greatest eaters I ever saw.

My four couple of fox-hounds are reduced to a single & some of them
having lost themselves, others [have died] but the couple of blood-hounds
are increased to five couple. I set up my [?] lately & have got two pair of
as good horses as my indian friend: they will draw a greater weight than
his, but will not travel so fast. Next year I shall have a strong set when
all the ten are in, & if I then come upon the fresh slot of deer shall ride
a-hunting most merrily & always be close in with the hounds. I walked to
Chatteau in March when the sun gave a tollerable degree of heat but the
whole country being covered with glazed snow the reflection made me go
stone blind the second day after my arrival there & the whole time I was
in the greatest agonies, the feel being exactly as if my eyes had been full
of scotch snuff. Many people have entirely lost their sight by it, but that
I suppose owing to improper applications. The prescriptions of those I
consulted were so contrary to my own judgement that I determined to
doctor myself & accordingly [?] better things fomented them with water
only which restored me to sight the second night tho' I was told not to
expect to see in less than a week, perhaps a month. As prevention is
always better than a cure, I invented a method of preventing the like accident happening again, and have ever since gone out all weather without

damage; yet I could not prevail on my blockheads to do the same & they have all of them been blind over & over again. I refer you to my father for further particulars & am Dear Sir ... etc.etc.

From George Cartwright to William Cartwright Esq.[23]

Honour'd Sir   Ranger Lodge 2nd June 1771

I have once more the pleasure of enquiring after your health & that of the rest of the family at Marnham, and acquainting you that I am still alive and well. At the beginning of the winter I was in a very critical situation. Mr. Lucas [arrived] on the eleventh of October with eight hands & about a month's provision: immediately after the weather changed apace & the river began to freeze. In a few days a Sealing Crew of ours arrived but brought me no letter or any account of my brother's arrival at Fogo, but on the 21st a Shallop came with provisions which relieved me from my anxiety. We had very fine weather afterwards till the beginning of December, when there were a few sharp days; but upon the whole the winter was very good. Whilst my Thermometer was in being it never froze [rose?] more than 52 degrees. It was unfortunately broke in December which prevented my making any more observations on the severity of the frost, which, I think was greatest in January. It several times brought into my mind those lines in Hudibras
 "And many dangers shall environ
 The man that meddles with cold iron" — for whenever anyone forgot himself in setting a Trap & laid hold of it with his bare hand, it immediately stuck to it like sealing wax to paper.
 Mr Coghlan sent me off two other Sealing Crews (They all consisted of six hands each) but they settled upon the Newfoundland shore & what they did I have not yet heard. The one here killed 800 Seals (which we judge will produce 14 Tons of Oil) besides about 58 foxes. The people I had in the River made but bad work of it. Two of them with Charles & myself attended the furring. We had but about two dozen of traps and most of them bad ones (our new ones did not come from England) besides the two furriers here were but ignorant men. Every body who had lived here before insisted upon it that there were no Beaver; but from what I saw in

the winter I was positive to the contrary and after an infinity of pains was
confirmed in my opinion, by finding two new & above fifty old houses;
but that not being till after the ponds were all fast & having no dog that
would find them under the banks when they lay out, we killed but five;
three of which were Wilkites weighing 45 lb each, the other two 14. They
are most delicious eating when fat; at other times but indifferent. In the
latter end of October I had the luck to kill a very large stag, & a fortnight
after, standing at the door I saw fourteen coming down the river upon
the ice. I snatched up a rifle & should have had an exceeding fine shot at
a good hind, but my gun was not loaded & away they scampered into the
woods. However clapping a couple of rifles upon my shoulders I pursued
them and in about three hours came up with & killed a brace of them. My
larder was then most luxuriously furnished, that with oeconomy & some
little help from the beaver, growse, hawks & ravens we did not finish the
last of the venison till the 17th of February.

The river was froze up from the 2nd Novem^r. to the 13th May: the still
ponds are not yet all broke up & there is now ice before the door above
three feet thick, although it lies in a warm cove full to the sun. I had my
health very well till about the middle of January when I was very indifferent for about a month, owing to a cold I got by lying in the woods four
nights following in very severe weather. During my indisposition a very
melancholic accident happened. Mr. Jones the Surgeon of the Garrison
at Chatteau coming over here imprudently by himself lost his way & was
froze to death. I did not know he was missing till four days after, when my
people searched and found him upon Round Island. He had got within
an hour's walk of this house when he unfortunately took my footing upon
a pond which lies on the Neck between this River and Niger Sound &
followed it. I happened to go that day over to the Sounds to look at a river
which lies in the s.w. corner of it: after which keeping down the south
side, crossed it towards the bottom and went at night to Seal Island.
[?] not only because he was a good man, & my being innocently the cause
of it, as he was coming to this house: but it was very near making me
accessory to the death of two more, for he was coming to deliver my maid,
and his loss together with the want of better assistance obliged me to take
upon myself the office of Doctor. My trouble did not end there, for I was
constrained to officiate as nurse next & take the child to bed with me, her
husband having enough to do to attend her. [lines struck out] She would

have done exceedingly well but for her own imprudence in getting up
to soon, which gave her cold & obliged me to transmogrify myself into
a physician and an apothecary. Here I had like to have failed for want of
tools for not one gram of what was wanted did my shop contain. The poor
woman grew worse, was in imminent danger of her life & I found it was
absolutely necessary something should be done, kill or cure, therefore
carefully perusing my friend Dr. Brooks, & considering the qualities
of each sample concluded that a plant which grew in great abundance
hereabouts was of the same kind as some of those prescribed, gave her a
strong decoction [?].

I fear I've dwelt too long on this subject, but hope the novelty of it will
plead my excuse. My reputation as a Surgeon too is perfectly established
on this coast; I have healed the wounds & shed the blood of most of the
people hereabouts & not one of them but declares I handle the lancet
better than any surgeon they ever felt: in short I have had a deal of practice both in physick and surgery & every one of my patients have done
well which is more than the best physician or surgeon in London can say.
But let us return to business. I have cut down more trees on my estate
this winter than many men of fortune in England have on theirs. I kept
two saws going most of the time, but the stuff being bad & the work men
worse, we have not a great deal of board to show for it: how ever we have
built two vats & cut stuff enough to build a salmon house, covered wharf,
& a stage for the Cod fishery. Our boats are repaired & one third of the
salmon house is up. The quantity of fur but small, it consists of Marten,
Fox, Squirrel, Otter & Beaver. We have killed three beasts of a kind
we are not acquainted with — I am of opinion they are sables, and one
Carcazew.[24] It is of a very mischievous nature & in shape partakes with
bear & wolf, but not so large as either: his skin of little value. The cursed
wolves did me a good deal of mischief. They carried off two traps & left
some toes in two others. Our Sealing Crews also lost two traps by them,
& a third was near gone but they came up just in time enough to shoot the
wolf: He was a brave stout old beast but not very well tasted. They sent me
a quarter, but 'twas as much as ever I could do to prevail on my stomach
to receive it. You will wonder that we got but little fur but the three best
months in the year were elapsed before the traps or furriers arrived, &
the fourth was spent in rummaging the country. But I know it now pretty
well and am thoroughly convinced there is a vast fortune to be made here

with industry in a short time, & I have got insight enough into the business to know how to make it too & will before I leave it: yet I shall not be at all surprized if we lose money as we at present go on. This goes to Newfoundland in a shallop which I now dispatch for oil casks etc. & I hope her return will bring me letters from England. I am ... Honoured Sir ...

From George Cartwright to Mrs. Cartwright (probably his mother, Ann Cartwright)

Honour'd Madam   Chatteau 24th August 1771

During the course of this summer I received several very acceptable letters from Marnham, for which I stand much indebted to the respective authors. I can not expect credit when I declare I have not yet had time to answer them particularly, yet 'tis a fact; nay I can hardly find time to shave myself, but as soon as I have dispatched our little Nimrod[25] (Watson, Master) shall have some leisure, when my father shall have all the information relative to the trade we are concerned in, as far as my abilities will allow; and an opportunity of writing again will soon offer. Through the means of the Indian family which wintered with me my interest is pretty well established amongst those savages. Twelve shallops of them came here this summer and called on me in their way when I joined the squadron and moved with them to Cape Charles, Camp Islands & this harbour, pitching my tent amongst theirs. With only one of my [line obliterated by paper tearing] day for my tent was full of them from morning till night. I depended on the great provisions of friendship which they had made and was not deceived; for they behaved very well and did nothing without my consent. The Chief lately accompanied me to Fogo. I shall have the agreable company of another family or two next winter and intend not to be without them. I have made great progress in their language, which is a more uncouth one than Malabar. We have had an exceeding bad summer, which has hurt the cod fishing much by constant gales of wind; and more than once they had like to have set me diving for muscles.

Grants will soon be given upon this coast for the further encouragement of the fisheries; which indeed cannot be carried on to the greatest national advantage until the adventurers have some better security for risking their property than they enjoy at present. I wish my father could procure me one of as much extent as I could occupy and in such a situation as my brother can point out to him; for we reconnoitred the coast together last year.

From George Cartwright to William Cartwright Esq.

Honour'd Sir   Ranger Lodge [paper tearing] [Sept. 1771]

As my mother acquainted me that you was desirous of knowing the advantages to be made from the several occupations upon this coast, I shall now endeavour to lay them before you.

The Cod fishery is a considerable branch, but much more precarious here than on the Newfoundland shore. It requires great knowledge, a large capital; a vast number of hands & boats to carry it on, so that I shall say no more on that head. The Salmon fishery may be followed in many different places, for all the Rivers we yet know of abound with this fish. Three men, three nets with two small punts are in general sufficient for one river. The season lasts from the latter end of June to the beginning of August. What fish is caught later serves for winter provision. As soon as the salmon is packed the men repair the buildings, put out traps for Otters, Slips for Deer & nets for Bay-seals till the middle of September, when, being all Furriers they betake themselves to the woods for the winter, where they kill Beaver, Otter, Marten, Deer, Bear, Wolf & Fox. In April they return but put out Otter & Fox traps along-shore. When the ice breaks up they set nets for young Seals, kill Ducks & gather eggs, of which we got about a thousand last spring, and might have got three times the number had we had time. In May they are getting Rinds for roofing the Salmon houses & other buildings and put the Cribs and every thing else in order for the ensuing season, by which you see they do not want employment.

The Sealing is a profitable branch of business. In the beginning of October the Crews should repair to their Posts & get every thing in order. The Seals come from the northward about the second week in November & are all gone by before the middle of December. When the work is over

the nets are hung in a house to thaw & dry; the seals are covered with
snow to prevent the frost from turning them, which it otherwise would do
equal to fire. Now they put out traps for fur near the post & take advantage
of every thaw to skin out & put the fat into the vats or casks till spring.
But in May is the great work of skimming & melting out the oil. The casks
are [?] in the shade to prevent the heat of the sun from causing a leakage
in things. The nets are again dry'd more effectually, the skiffs lodged in
safety and the people having delivered up all their craft are dismissed to
follow their other occupation of cod fishing for the summer.

As for the Indian trade it will be carried on to most advantage by the
Moravians, or rather monopolized by them, in consequence of their
being settled in the midst of the savages by means of a Grant from the
Government. I have some trifling traffick with them when they were in
this neighbourhood, but it can no longer answer to pursue it with a vessel
fitted out at a great expense as we did last year. Beads are the staple commodity in this barter, after the indians have supplied their wants in Tools
of iron & a few other conveniences; and in return they furnish whale-
bone, young Seal skins, Fox, Otter, Martin, Deer, Bear & Wolf skins in
small quantities.

From George Cartwright to Lieut. John Cartwright

Ranger Lodge, 20 Sep. 1771

Four days after you sailed, the Enterprise returned. Poor Lucas brought
an indian family with him, which wintered with me. Soon after a Crew
of Share-men arrived at Seal island. They killed 800 Seals, 33 Foxes,
5 Otters, 1 Deer & 1 Wolf. On the 21st October after dispairing of supplies the boat which Coghlan sent in consequence of your arrival at Fogo
arrived with Provisions & Winter Crews. Two men with Charles & myself
commenced Furriers. I was the best of the crew, so you may imagine we
made a fine hand of it. But I will pass over the affairs of the winter, having
filled my other letters with them, & come to occurrences of the summer.
On the 4th of July five indian shallops arrived at the Cape. Some of the
men came up here that night & lay in the dining room. Next day I went

with them to visit their wives; the day after the women & children with
some men came up here again, 32 in number; a goodly company! They
were exceedingly pleased with their treatment, delivered a great quantity
of salmon & carried more away with them. At night five did not choose to
leave me & so they spent their evening here. The next day I took Charles
with me & pitched my tent amongst them by the side of the little tickle.
My winter friends had made so favourable a report of me that they all
behaved exceedingly well, nor did they steal anything although the tent
was full from morn till night & goods were lying about. I employed my
time so well that before it was known at Chatteau that they were here, I
had got almost all the things they brought for truck but they amounted
indeed to a very small quantity as the Moravians take care to supply all
their wants before they leave home. On the 16th I accompanied them to
Camp Islands & took my passage in the boat of their Chief; after a stay of
two days we proceeded together to Table Point where I killed a brace of
Deer. On the 26 five boats more came, four of which went directly to the
rest, the other came up here. They not having seen those who wintered
here began to behave in their old way, and presumed on their numbers
being no less than thirty five. I had only two men at the house but I soon
convinced them they were wrong; for one behaving with great rudeness
to Nanny I seized him by the collar and gave him such a launch out of the
door he was near going head foremost into the river, then made him get
into his canoe & march off, and ordered the rest to take themselves away,
which they did immediately. Two days after I went to Camp Islands again
but their last boats had met with the Moravian Brig to the northward &
sold almost every thing: August 4th I sailed for Fogo taking with me the
Chief; on the 20th we returned to Chatteau when I found that the indians were gone to Quirpon & up the Straights. On the 1st September they
returned, the Chief joined them & early the next morning they went off
home, carrying with them two baskets besides some seal-skins & whale
bone I had given them to build me an open canoe with. I speak the language tollerably well & had the two families wintered here according to
their promise; should almost have been master of it by next spring.

I am now more than ever of opinion that Lucas was very partial to these
people; for I think them but a sorry race of mortals, with but little of
honour, honesty, gratitude, fidelity, or any one natural virtue to recom-

mend them; while they are riplete with revenges, fraud, selfishness, avarice and every vice, except drunkenness, I almoſt ever heard of. We cannot expeɗ they should behave otherwise to us when they rob and murther each other, are governed by no laws but their own will, which prompts them to revenge with the knife or arrow every slight affront or injury. Thanks to Heaven they are great cowards! What some have asserted of the chaſtity of their women is moſt absurd. Not one of them will refuse her compliance for a pennyworth of beads if an opportunity offers; nor a man will refuse you his wife for a night for a few ſtrings; but I think the payment should be on their side, for they are far more filthy than swine [paper tear] persuade myself that I am looking upon women. I had plenty of offers from both the men & their wives, but did not choose to part with my goods at such a rate. Nothing is more common than an exchange of wives; and a dog has nearly as great an idea of modeſty as they.

Lately I have had several good dishes of green pease and in winter could have set out a table of five & five without salt meat from Oɗober to March; besides a Dessert of Jellies, Cuſtard & Tarts etc. etc. etc. I had one Beaver house in the Upper Island Pond & another about five miles NW of the House. We found above fifty old ones & none of them [were] furred: it was late before we began to rummage & snow soon coming on obliged us to desiſt. The Deer beat to the southward by us; then weſterly along the Straights from the middle of December to the middle of January; crossing the South Head, or Punt Pond on their way to Niger Sound, thence by the head of St. Peter's and Temple Bays they get upon the weſtern ground. Between Rack Cove and Bradore they are in the greateſt abundance till the beginning of April, when they return and it is June before they are all [paper damage] then ſtraight into White Bear Sound & Flat Point; and after the ice breaks Flat Point & [paper tear] are their principal crossing places. Notwithſtanding what we have [believed we] found them ten times more difficult to get a shot at whilſt the snow was on the ground than in summer; for then they always frequent the high barren hills and keep so good a look-out, and are so shy, 'tis next to an impossibility to get near them, besides they rove about so much and make such short halts, that there is little chance of meeting with them. I have slotted them for days together without coming up with them, although I have seen them at going off. They generally travel down wind, as their nose then informs them of any pursuit. The only chance is to forelay them on the march; but we never had an opportunity of doing so except once in May, when one of the Furriers killed a hind: but she was so rotten and poor that we could not eat a bit of her.

My excursions this winter will not be so fatiguing as last, for I have got an indian slide to be drawn by dogs. The bloodhounds are increased to five couple and all broke to the harness so that when I get upon the fresh slot of deer I shall travel at a merry pace. The Table Land is the best place for Deer and White Bear in Spring of any I yet know. I shall live there from February till June, burrowing like a rabbit in the snow. I must get you to order me a tent from the Oil Cloth Manufactory at Knightsbridge; one side white the other brown; it must be the shape of a soldier's but supported by a strong wooden frame as well as poles, for in winter it will be covered with snow, in summer with boughs or sods. Let the frame be 8 feet long x 7 broad; joined at the angles with screws. The ridge pole to be 5 feet high supported at each end by a perpendicular pole: & from the ridge pole to the frame on each side have nine rafters, (pointed with iron at each end) for bearing up the snow etc. The tent to be in one piece & closed at each end; except a hole 2 feet square by the side of one of the perpendiculars over which must hang down a canvas door. The holes in the ridge pole & the frame, which receive the iron points of the rafters, must also be iron bound: on the outside too of the ground frame all round, let there be short stumpy spikes of iron; corresponding with eyelet holes in the border of the tent in order to keep it tight & close down. Pack all in a chest with [paper tear] and spare canvas for repairs. If it is well executed I shall be able to shift my winter quarters, as often as I shall be inclined.

Letter to George Cartwright from Daniel Sutton

Sutton House, London, Jan. 10, 1774

Sir

When I had the pleasure of seeing you last little did I expect there would be occasion for troubling you with this address. The Day after you left me Noozelliack was attacked with the usual Symptoms proceeding the Eruption of a benign Small Pox. But he apprehending they were the consequences of the medicines & not any particular Disease coud by no means be prevail'd to take any one thing either medicine or food, cold water excepted, in this way he continued for three Days after which time the Small Pox made its appearance not more in number than 2 or 3 hundred. He had had a slight cough from his first coming hither which at the Eruption of the Small Pox increased to violent degree. The next Day the Cough not abating I sent for Dr. James who ordered him to be blooded & if possible give his Powder. The former the Boy complyed with but the powder he could by no means be prevailed upon to take continuing in his obstinacy & determined to take nothing but water he fell a Sacrafice this morning about 3 o'clock.

I ransacked London all over for a person who coud spake his language but coud not procure one that he could or wd. understand.

Apprehending a speedy Dessolution I found I had not time to get a letter to you soon enough for you to be of any service.

I would have observed some time before he died he was attacked with an incessant vomiting. Upon the whole I apprehend this fatal catastrophe was entirely owing to an invincible aversion to all kinds of food or medicine from the first approach of illness & my not being able to prevail upon him to take what was absolutely necessary & what in all human possibility wd. have carried him through. I have given orders for his decent interment and am Sir your very unhappy Servt.

Danl. Sutton

Letter from John Cartwright to Catherine Cartwright (sister of George and John)

My Dear Miss Kitty

That I deserve a great deal more than I suffer for my late error I am ready to acknowledge; but dont you insult me, Madam! with your plaguing prudence. I am sufficiently haunted with self-reproaching reflections, and with dreams of recommitting the same folly, so spare, I pray ye, your additional mortifications.

But you seem to be in a mistake about the matter. I was not so wild
as to direct to Mrs. Close intentionally, but her name for ever ran in my
mind when I meant to write Rice; and I had very nearly done the same
thing several times before. I understood our friend in her last that for
the present she would get her [?] Charles to correspond with you but you
seem to say that all correspondence whatever is to be suspended. This I
suppose is likewise taking up a prudent resolution! I shall, at this rate,
soon grow out of conceit with all prudence what ever, but I suppose I must
be all patience on this occasion as having been the offender. Well! be it
so, as long as I can hold out. If you should be so happy as to see Miss Stone
remember me to her in the strongest terms. I shall regret my absence
from Marnham on that occasion. We hard drinkers go on extremely
well. My sister is very near as gay as her dear Cornaro was at ninety, for
I presume she must haunt her prayers since she is singing all day long.
This morning I went into the Queen's Bath which I mean to repeat twice
or thrice more for a few minutes each time.

Dr. Harrington intimates to me that I must not be surprized if this and
my internal ablutions together should bring on another touch of the gout,
but I hope my trotting horse will preserve me. Should I entirely escape,
I shall hope that another two months drinking, after an intermission of
the month of March, will perfectly restore my health. It is in April & May
that the waters are chiefly recommended and I find Harrington quite in
this opinion. Indeed I have not been nearly so sensible of good effects
from the waters since I have been here as from riding exercise. I never
fail to feel the most beneficial effects from it, but when very bad weather
prevents my riding I am scarcely sensible of the efficacy of the water. It is
with the most scrupulous attention to my eating that I keep my stomach
free from disorder, and very rarely indeed find the waters create that keen
appetite they did last year. But nevertheless, I gain strength, spirits &
bloom so that I am content. My stomach has been so long ill & weak that I
do not expect it to be well until a regular & long practiced temperance has
recovered it.

We have had Scot here for a few days. His friend Elphinstone remains
and gives him an extraordinary good character. In the pump room the day
after Scot left Bath he sayd that was the first time they had been parted a
day for many months past and seem'd to express himself as if it gave him
much concern. This will scarcely find George at Marnham as the time

of his departure is now approaching & he had a good deal to do. If he has not left you tell him that Shuldham is gone to town but that I saw him & delivered his message. I beg to know the [day he] himself means to be in London. Tell him [that] the well known Commodore Walker (in partnership with a Mr. Bailey of London) is obliged after a long perseverance & great loss to drop the Codfishery in Chaleur Bay, Nova Scotia, & confine himself altogether to Salmon fishing.

With Duty etc. to all at home believe me

Dear Kitty
Your affect. brot. John

22nd Jan. 1774

Letter from the merchant firm H. Hunt regarding Cartwright's Sandwich Bay assets

24 Dec. 1818
Hen[y]. Hunt

Dear Sir

Accept my thanks for the information about Sandwich Bay which Capt. Cartwright has obliged me with. Agreeable to his wish I now return you his copy of the Proclamation with all the information I have on the questions he asks.

Philip Beard and Co. paid Noble & Co. for the posts & Goods remaining including some Salmon £2710 in which was included 480 Tierces of Salmon and the usual Netts & property remaining after a fishing season. The posts I think were valued at £300.

Dear Sir,
[?] for
Henry Hunt
24 Dec. 1818

Total Cost   2710

480 Tierces Salmon w. 40/	960
	1750
Posts	300
Nets, etc.	1450

Letter from Hudson's Bay Company to George Cartwright, followed by a note resembling Cartwright's handwriting

Hudsons Bay House

11th June 1818

Sir

I am directed by the Governors and Committee of the Hudsons Bay Company to request you will have the goodness to communicate with Mr. Blagg & to transmit to the Governor the terms upon which he will be disposed to engage with the Company as the Chief of an Expedition whose object will be to establish a trade with the Esquimaux particularly whether he will be inclined to engage upon the same terms as the other officers in the Service. These are paid by a Share of the Net profits of the Trade and the rate of that Share must be a matter of future arrangement.

You will also have the goodness to say upon what terms you will part with your Book of remarks upon Labrador.

I am, Sir, your most obed't hble Servt.

Wm. Smith, Ass^t. Sec^y.

[note in handwriting resembling George Cartwright's] N.B. If W.B. is disposed to go to Hudson's Bay he must procure leave to do so from the Commander in Chief, which will not be refused, and he must enquire before whom he can make his affidavit for Half-pay. If the Governor of the H.B.C. makes an application to the Lords Commissioners of H.M. Treasury, they will most probably make that matter easy. In Labrador, Capt. Cartwright used to make his before the Master of one of his Vessels, who signed himself Fishing Admiral, and they were always regularly paid, but there are no Fishing Admirals in Hudson's Bay. The Cashier of Half-pay should be consulted in the first instance if the Commander in Chief's leave has been obtained.

Four letters concerning the placement of the memorial to George and John Cartwright in Cartwright, Labrador, and two letters concerning a memorandum [possibly "Memorandum for W"] written by George Cartwright, which was loaned to Dr Grenfell in the 1900s

From Frances Dorothy Cartwright to George Cartwright, possibly her son

10 June 1858

My very dear George – I feel quite ashamed of not having sooner explained to you what you ought to know & in which you take so much interest but really till within these few days, I have been too weak, and to write to anyone. Today, I am happy to tell you, I have got my steam up a little & hope with God's blessing (without which I know we can do nothing of ourselves) to live till our Labrador matters are in good train. In the first place I received from Dr. Field [*sic*], Bishop of Newfoundland, a letter dated Halifax, Nova Scotia 1ˢᵗ May 1858 – having received mine at Bermuda – just before he left it. Nothing can express more kindness, I might say, friendship, he anxiously wishes for a church to be built besides the memorial which he highly approves of and he gave me at the same time an introduction to some London Merchants, concerned in the fisheries to express his wish that a small Church should be built to which he wd. furnish the fittings. I wrote accordingly and was ansd. very politely, but was referred to another merchant for <u>his</u> opinion etc. Well, as you may suppose these fishery gentlemen want no Church nor even a place of worship on the coast and consider it not at all <u>requisite</u> or <u>practical</u>. But as Fanny Strickland said furiously to me "Aunt, you are Cartwright to the back bone." So I wrote to the Bishop and told him that it might still be practicable to build a small wooden Church which is the material most recommended in that climate and for which his Lordship had so generously offered the fittings and that it might be adjoining the Cemetery already consecrated. It is objected to, that it was not the <u>exact</u> spot on which Capᵗ Cartwright lived, which was farther north on Marnham harbour but what does that signify; & the name is enough and as I took the liberty of suggesting to the Bishop it may, <u>please God</u>, be a nucleus for future benefits to those persecuted Exquimau [*sic*] who seem to retain the tradition of former friend & Protectors of that name.

I send you a copy of the intended inscription[26] of which the good Bishop highly approves — You muſt observe that in joining the good old Major's name to the Memorial, it is only juſtice, to one who was [act] and part with his elder brother whom he assiſted in the Labrador researches, and who as surrogate or judge wrote such a spirited remonſtrance to the governor of Newfoundland, that the poor natives were no longer hunted like wild beaſts; moreover, though he punished those fishermen who were guilty of cruelty to either in Labrador or Newfoundland, not one of his judgements were ever disputed or reversed in England. It now remains to say that I am not without hopes that all may be managed without any great difficulty, when I receive an ansr. from the Bishop, to recommend me in what way to proceed as soon as I know with some degree of certainty what would be the expense of such a ſtructure erected in the Spot etc. Meanwhile, I shall be collecting all my little savings, certainly all my <u>flounces</u> & practising for the firſt time in my life the moſt rigid economy. Now don't you laugh & quiz your poor old Aunt, you bad man, if you should see her going about in a common ſtraw bonnet, and brown holland cloak, not forgetting coarse worſted ſtockings!!!

I suppose if I were to come to Lyme, your Worship would take me up for a Wagabond. But a truce to nonsense — I think I told you W. Strickland will give me £30 in advance & will kindly assiſt me afterwards. However I am afraid all this will be a little premature, as it muſt be some weeks before I hear from his Lordship. Meanwhile I know I have your good wishes as well as Henrie's & remain as ever

Your affect. F.D.C.

I have in my possession a moſt intereſting pamphlet written on the subject of those coaſts by my Uncle <u>the Major</u> when a Lieut. on board the Guernsey Man of War in 1768.

How graphic your description is of your Labrador experiences!

I am much obliged by your mentioning Dr. Livingſton's Book, which I will read as soon as possible. Helena proposes to come on the 15 & ſtaying till the 26 July. Please to return me the copy of the Inscription as I have no other.

A letter, author unknown, containing a brief reference to Captain George Cart-wright (letter illegible and not transcribed in its entirety)

Cliff Cottage
March 20, 1861

My dearest Georgy

It was really kind of you writing the nice little Sunday afternoon note, and now, I rejoice to see that you really seem [?] deriving benefits from the diversity you are according your kind & hospitable friends; do inquire who the Mr. Cartwright is still at Nottm. Very probably a distant con-nexion, though Aunt Frances did not seem to know of him. I am reading Mr. Howitt's charming book on the Forests of England; it was published in 1844, he says of Sherwood "There twenty years ago Captn. Cartwright of Labrador might be seen, hunting with a fine pack of Dogs." I enclose all the notes [?] burn them, when read. John is an excellent Father, & a brick [?] fellow. I am, thank God, remarkably well, today is brilliantly cold ...

From George Goodridge, Cartwright, Labrador, to Bishop Feild

Cartwright 10 Oct. 1861

Sir,

Bishop Feild,

[Illegible opening lines] ... Church, copy of which I have sent to Mr. C. Hunt, the people cutting the framework here & some of the board & plank, triple more [?] would complete it. I shall frame it so Mr. C. Hunt [?] Church being built here.

There was difficulty in erecting the Tomb, as all the pieces were fitted, & the copper clamps I hope were also sent with the Tomb; greatest dif-ficulty we had of getting it to the Cemetry, some of the pieces were very heavy & had to be hauled a [*sic*] Coast through a deep bog on planks. It has fine appearance. I don't think there is nicer looking Tomb in any of the Cemeteries in Nfld. Since it has been erected I am making road which is nearly completed from our Dwelling House to the Cemetry. In case a Church should be built it would make a good carriage road but such a

Vehicle I expect will never be seen on this coast. The officers of the Man of War, Doctor in particular, were much pleased to visit the Tomb enclosure in the [aft.]. I paid for erecting it. Have not arranged anything for labour. I have credited for the Biblical Testaments I [?] some two or three years since & have few left which will be sold this winter.

I have sent a small case of preserved Salmon to care of Messr. R. Prowse & Sons.

We have had a very poor fishery but not so bad as it has been at other places. I leave in few days for Grady's Isl. I [?] from what I can learn I am fearful there will be some disturbances again this fall when the Labrador men return.

[Last lines of letter are illegible].

[Signed]

Geo. Goodridge

Unsigned letter, probably from Bishop Feild. At the bottom of the first page, it is addressed to "George Cartwright Esq."

St. John's Nf.

5 Nov. 1861

My dear Sir,

I am happy to inform you that the Monument has been successfully erected, in the place, at Cartwright Harbor, originally chosen by Miss Cartwright. Dr. Lillburn of H.M. Ship Hyona has seen it and speaks very favorably of its position and appearance. I enclose a letter I have just received from Mr. Goodridge which supplies particular information in reference to the work: Mr. Goodridge is the Agent of Messr. Hunt & Co. and it is due to the interest he has taken in the work that it has been brought to a satisfactory accomplishment. We are very much indebted to him and you will see by his letter that while he would do just honor to the dead, he regards the higher interests of his living neighbours, the children descendants (many of them) of those whom Captain Cartwright first raised from the state of savages. I think a line of acknowledgement from yourself would be very gratifying to him, and his services deserve such notice.

Should you have any occasion to write to me, please address me (till Easter) at Bermuda.

I should be glad to know that Miss Cartwright is in a state capable of understanding how far her affectionate wishes have been carried into effect.

I am, my dear Sir,
Yours faithfully.
[?] Newfoundland

Typewritten letter bearing the letterhead of the Royal National Mission to Deep Sea Fishermen

Francis H. Wood, Secretary
Bridge House, 181 Queen Victoria Street,
London, E.C.

April 20th 1914.

Dear Sir,

I understand you were kind enough to lend the enclosed letter to Dr. Grenfell, and he has asked me to return it to you with many thanks for the loan of it.

Believe me to be,
Yours truly,
Frances H. Wood
Secretary

George Cartwright Esq.

[handwritten notation at bottom of page "Acknowledged 23.4.14" and "PTO"]

[This letter was enclosed with the above]

21 Old Sq.
Lincoln's Inn
6 May 1914

Dear Sir

Do you happen to know if Dr. Grenfell has finished with the Diary of my Great Aunt Catherine Cartwright which I lent to him at the same time as the memorandum you kindly returned to me last month.

I don't want to hurry Dr. Grenfell, but should you see him will you please mention the matter to him.

Believe me

Yours faithfully

Geo. Cartwright

F.H. Wood Esq.

The unnumbered loose papers found in the Cartwright archive are unrelated to one another, but each stands alone as a document of some historic interest, which by its contents can be tied to Labrador matters.

A table of itemized expenses jotted into a narrow 8.5 by 24 cm leaflet, dating to 1790–91, lists the costs of producing the Journal, *published in 1792. A map, undoubtedly of the Makkovik area, bears an explanatory note in Cartwright's handwriting, "A Sketch of a Bay in Labrador copied from one drawn by an Eskimeau Indian." A list on a 6.5 by 17 cm scrap of paper, entitled "Provisions for one Man for Twelve Months," gives the costs of food, goods including "Goods for Ind'n. Trade," and freight. Another list on a scrap of 7 by 18 cm paper enumerates employees "At Home" and elsewhere. A 5 by 10 cm bit of paper lists the wages of each employee type alongside costs of provisions, furniture, shipping, and other items. The sixth item is an 8 by 18 cm piece of paper bearing a table of employee numbers and their tasks against their locations at "Home," "Seal Island," "Venison Harbour," and elsewhere.*

Expenses attending the Publication of my Journal[27]

1790		
Jan. 1 to 21	Letters sent	" – " – 5
	Letters received	" – " – 7
	A quire of Paper	" – " – 1
23[d.]	1 L[re] rec'd.	" – " – 1
	1 d. sent	" – " – 1
	Expenses at Newark	" – " –
25	Letters	" – " – 3
26	Letters	" – " – 9

27 [? ink stain]	Parcel rec^{d.}	" - " - 2
30	D^{o.} Sent	" - " - 2
31	Letter sent	" - " - 1
Feb.		" - " -
6th	Letters & Parcels	" - " - 4
9	Letter	" - " - 1
13	d^{o.}	" - " - 1
14	d^{o.}	" - " - 1
15	d^{o.}	" - " - 4
18	Expenses at Worksop	" -17- 3
19	Letter rec'd.	" - " - 1
20	D^{o.} Sent	" - " - 2
23 & 4	d^o rec'd & sent	" - " - 2
Mar 3^d	Ltre. rec'd.	" - " - 5
7th	Letters	" - " - 6
16	d^{o.} rec'd.	" - " - 11
20	d^{o.} Sent	" - " - 1
Apl.		
18	Do. Do.	" - " - 1
June		
2	Do. rec.d	" - " - 6
4	d^{o.} sent.	" - " - 2
13	d^{o.} rec'd.	" - " - 6
July		
12	1 d^{o.} sent	" - " - 1
14	1 d^{o.} rec'd	" - " - 5
15	1 d. sent	" - " - 1
17	1 d^{o.} rec'd	" - " - 4
18	1 d^{o.} sent	" - " - 1
29	Parcel	" - " - 2
30	Letter, Parcel	" - " - 4
31	Parcel	" - " - 2
	Carry over	1 - 6 - 1
		£ s d
[new page in original]	Brought over	1 - 6 - 1
Aug. 6	Letter rec.d.	" - " - 6
	d^{o.} sent	" - " - 1

	Parcel	" - " - 2
10	Letter rec'd.	" - " - 1
22	1 d°· sent	" - " - 1
25	1 d°· d°·	" - " - 1
28	1 d°· rec'd.	" - " - 6
29	Parcel	" - " - 2
Sept. 4	2 Letters	" - 1 - 4
5	1 d°·	[ink stain]
11	letter rec'd	" - " 9
15	2 letters sent	" - " - 2
17	3 d°· rec'd.	" - " - 6
18	d°· rec'd.	" - " - 2
19	d°·	" - 1 - 6
20	[?] d°· sent	" - " - 3
	1 d°· rec'd/	" - " - 2
21	1 d°· d°· 1 d°· sent	" - " - 2
22	2 d°· sent	" - " - 2
23	2 d°· sent	" - " - 2
25	Letters	" - " - 4
27	d°· & Parcel	" - " - 3
28	Parcel	" - " - 2
Oct. 6	D°· to Hull	" - " - 10
	Letters	" - " - 10
Dec. 2	d°·	" - 1 - 4
14	1 d°·	" - " - 3
16	2 d°·	" - " - 3
20	2 d°·	" - " - 2
21	2 d°·	" - " - 2
28	1 d°·	" - " - 8
1791 Jany. 8	2 d°·	" - " - 2
25	—	" - " - 9
30	—	" - " - 1
Feby. 12	4 d°· rec'd.	" - " - 6
	3 sent	" - " - 3
13	1 d°· rec'd.	" - " - 6
14	1 d°· sent	" - " - 1
		£2 2 9

		£ s d
[new column in original]		
Brought over		2 - 2 - 9
Feby. 17	1 Ltr. rec'd.	" - " - 1
Mar. 3	1 d^{o.} rec'd.	" - " - 6
	2 d^{o.} sent	" - " - 2
5 to 17	d^{o.} rec'd.	" - 1 - "
21	d^{o.} rec'd.	" - " - 4
22	d^{o.} rec'd.	" - " - 6
27	3 d^{o.} sent	" - " - 3
to May 5th	Postage	" - 1 - 3
18	A Fox	" - 10- 6
20th	Paid Hilton	7 - 7 - "
June 14	Exp^s. to London	27-7 - 1
27	Letter	" - " - 1
July 16	Letters	" - 1 - 10
Aug. 18	Letters	" - 2 - 6
Sept. 25	Letters	" - 1 - 9
Oct. 11	d^{o.}	" - " - 11
28	d^{o.}	" - " - 11
Nov. 15	d^{o.}	" - " - 8
		£38 - " - 1

A Sketch of a Bay in Labrador copied from one drawn by an Eskimeau Indian

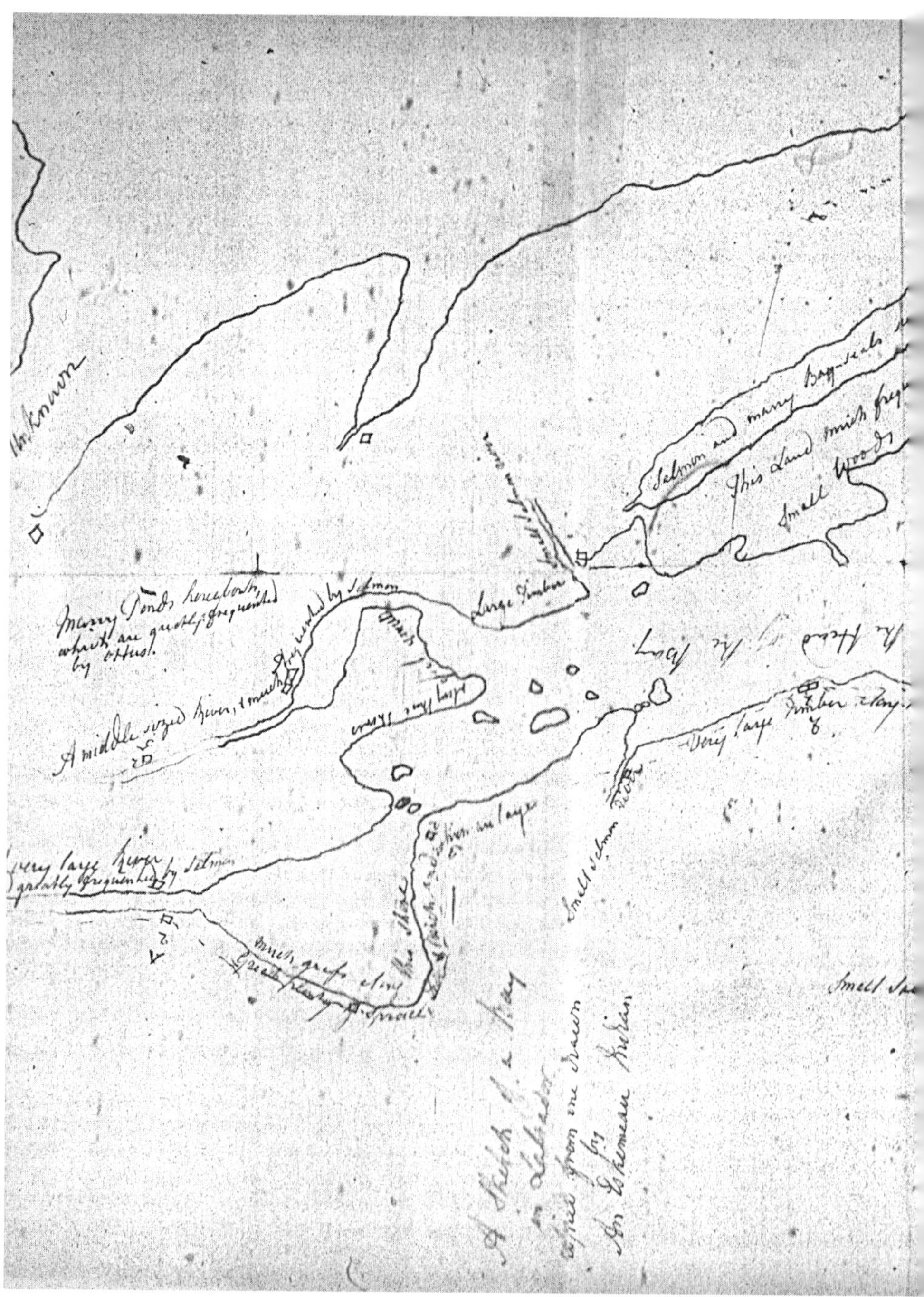

Abundance of
Islands lye off shore

High hills
and plenty
of white Foxes

in the Winter

Woods

Great plenty of Cod
found here all the Summer

well frequented by Deer in the
Summer

Tall woods of a tollerable size

Small Salmon river

Provisions for one Man for Twelve Months

[Probably based on direct experiences of Cartwright during his time in Labrador]

1 ½ Barrel of Pork
2 ½ Cwt. [hundredweight] Bread
2 ½ Cwt. Flour
19 ½ Gallon Pease
⅛ Cwt. Butter
3 Gallons Olive Oil
Ditto Treacle 6 Gals.
1 ½ Cwt. Dried Salt-fish (made there)

N.B. There should be fifteen Months Provisions laid in, lest an accident should happen to the ship in going out next Spring.

Provisions	500
Materials	500
Goods for Ind. Trade	300
Slops & Stores	200
	1500
Freight of vessel, 8 month at 50	400
	1900
Contingents	100
	2000

Employees "At Home" and Elsewhere

Furriers	8
Boatbuilders	2
Assistants to do.	2
Top Sawyers	2
Bottom do.	2
Squarer	1
Smith	1
Do. Assistant	1
Personal [sic]	1

Wooders	5
Hawk Bay N. Arm	2
Stoney Island	2
	29
Women	4
	33
Self	1
	34
Detached	12
At Home	22

Wages and Costs of Provisions and Other Items

Wages	£
1 Surgeon	60
3 Clerks	80
3 Boatbuilders	75
8 Coopers	160
1 Smith	20
1 Taylor	16
4 Sealers	146
6 Furriers	102
1 Joiner	25
2 Footmen	35
26 Youngsters	60
4 Women	32
1 Bricklayer	12
[Sub-total]	1023
Provisions	1950
Stores	2000
Furniture	2000
Shipping	6000
x	12973
Contingent	27
Total	13000

Employees at Various Labrador Locations

	At Home	Seal Island	Venison Harbour	Cat-Path	Hawke Island	Unappropriated	Total
Surgeon	1						1
Clerks	1	1	1				3
Boatbuilder	1	1	1				3
Coopers	2	3	3				8
Smiths	1						1
Taylors	1						1
Sealers		4	4				8
Furriers				5	1		6
Joiners	1						1
Footmen	2						2
Youngsters	6	5	5	5	1	4	26
Women	4						4
Bricklayer	1						1
Total	21	14	14	10	2	4	65

INTRODUCTION

1 Microfilm copies of the new Cartwright papers relating to Labrador are held at the Centre for Newfoundland Studies, Memorial University, St. John's, Newfoundland.

2 This form of citation, C.11.7.75, is used throughout the text for information taken from Cartwright's *Journal*. It follows Cartwright's daily journal entries, and is ordered as day, month, and year. For example, the "C" is for Cartwright and it is a *Journal* entry of 11 July 1775.

3 Moravian trade continued to be a point of contention for later merchants well into the 1800s, because the Moravians, who were wholly reliant on trade profits to keep their mission stations operational, offered better rates and were well stocked. The Hudson's Bay Company factor at Nachvak, for instance, was rankled by the 1873 price offered by the Ramah Moravian mission for a silver fox (one gun, one spy glass, and other items). He himself was sometimes forced to buy stock from the Ramah mission in order to keep Inuit at his post (HBC, Nachvak, Winter 1873).

4 Between the 1740s and the 1850s a number of names were given to Hamilton Inlet, including Baie des Kessessakiou, Baie de Kesselaki, La Baie des Esquimaux, Baie St Louis, Ivucktoke Bay, Eskimo Bay, Great Water Bay, Bouktok Inlet, Aivektok, Touktoke Bay, Ivertoke Inlet, and Grois Water Bay. In recent years toponymic confusion has been resolved at a cartographical level, with Hamilton Inlet referring to the eastern part of the waterway from Indian Harbour to the narrows at Rigolet. This is also called Groswater Bay. The western part is called Lake Melville. (See also Mailhot 2001, 2, for toponymic history of Kessessakiou).

5 In 1749, after Fornel, the Chateau Bay concession was granted to the doctor and naturalist Jean-François Gaultier (Gautico). In 1754 Gaultier deeded his rights to a Sieur de Lanaudière and Charles Gilbert (LAC-R11133-0-7-F, formerly MG8-F101; Roy 1940, 90; Roy 1942; White 1926, 9), which passed to British rule after 1763. Marsal's concession extended from Cape Charles to Alexis Bay. Altercations with

Inuit at Cape Charles resulted in deaths on both sides, and Marsal re-established himself further south at Petit Havre. Cape Charles was then leased to a Captain de Baune in 1749–50, who did not operate it. Marsal re-opened the post in 1754 until his death around 1756–57. The lease was then held by creditors until the time of British rule, but the post was not in operation (White 1926, 9; Roy 1940, 52, 107, 118; Roy 1942, 207; Anick 1976, 620).

6 An early petition to protest the British fishery system was tabled before Governor Murray in 1764 by William Brymner and seventeen Quebec-based traders (CO 42/1:219–21). Of particular issue was the loss of the lucrative seal fishery that took place each year in late autumn and spring, which was considered off-season under British rule. Another petition was submitted by Captain Nicholas Darby to Commodore Byron upon having his seal skins and oil confiscated for employing three Canadiens (CO 194/18:82–84). Darby must have been incensed at this measure against him because in 1762 he had petitioned on behalf of Bristol merchants for British control of Labrador in order to ensure a safer fishery (Crowhurst 1969). In 1771 John Noble and Andrew Pinson petitioned for the right to own property in order to carry out the cod fishery without sailing to Newfoundland to dry their fish (CO 194/18:86). Two further petitions were submitted to Governor Shuldham by George Cartwright and by John Noble and Andrew Pinson in 1773 (CO 194/31:5–15).

7 These portraits are part of a small collection of known images of Labrador Inuit who were taken to England. Watercolours of the heads of Attuiock and Caubvick were copied by an artist by the name of Hünneman in 1792 from Nathanial Dance's original full-figure pastels. The latter were in the collection of Sir Joseph Banks for many years and are presently in the collection of M.-J. Knatchbull. Although often unacknowledged, all published reproductions of the full-figure portraits originate from Lysaght (1971: facing p. 82). Banks sent the watercolour miniatures to his friend and colleague, the German anthropologist and anatomist Johann Friedrich Blumenbach. These portraits now form part of the Blumenbach collection at the Ethnological Museum of Göttingen, Germany. An unsigned portrait of another Inuit woman from Labrador, possibly Ickongoque or Caubvick, is in the Hunterian collection at the Royal College of Surgeons, London, England. Another portrait, of the Labrador Inuit woman Mikak and her son Tutauc, was painted by John Russell in the winter of 1768–69 for Joseph Banks. It was exhibited in London for a time and eventually presented to Blumenbach by Banks as a gift. It is presently in the collection of the Ethnological Museum of Göttingen. A family portrait was apparently painted of Attuiock and the other Inuit and was owned by Banks, but its whereabouts is unknown (Ethnological Museum of Göttingen; Lysaght 1971; Pearson 1978; Taylor 1983, 1984).

8   Cartwright was in England at the time of the Inuit visit to his settlement in 1782. References to a "boy" are made several times, for example C.20.6.79, C.22.6.79, C.18.9.83, C.19.9.93). The quote from the Nain Diary (3 October 1782) is:

> At night a Brother visited him [Tuglavina] in his Tent, where Abraham & Mikak resorted. They related that on their way to the South, they had been as far as Chateau Bay; and that the commanding officer, whom they looked upon as a Believer, had expressed his joy to see baptized Esquimeaux. He had told them that he was baptized and that he was not afraid to die. Tho' he was at that time very ill, he knew his Soul would go to Heaven, He also had given them his Proclamations of the Governor John Campbell etc. They had bought a large Boat, for which they were to bring payment next year, and Tuglavina and Abraham had each gotten a Gun, Shott, and powder. They had also been at a settlement of Mr. Cartwright; had seen his Son, who had expressed a great desire to pay us a visit."

9   Noble and Pinson remained the largest merchant operation in southern Labrador for a number of years, particularly after obtaining Cartwright's salmon, cod, and seal fisheries in the Sandwich Bay area in the winter of 1784–85. Their partnership dissolved in 1811–12 and the establishment was then taken over by Noble and Hunt. In 1816 it was sold to Philip Beard & Co. of Dartmouth, whose tenure there is uncertain. The firm of Hunt & Henley was established there by 1857 and eventually sold to the Hudson's Bay Company in 1873 (CO 194/62:176–183; DRO Lester and Garland Papers D365-letters; PC 1927, 1209; Anick 1976, 683; Marshall 1996; *DCB* for "Cartwright").

10   Sir Wilfred Thomason Grenfell (1865–1940) founded the Grenfell Mission, an arm of the Royal National Mission to Deep Sea Fishermen, which provided medical and social support to fishing families throughout northern Newfoundland and Labrador between 1892 and 1981 (Grenfell 1969; Rompkey 1991).

11   Some items in the loose pages have been omitted altogether because they repeat information in *Additions*, such as three sketches of hawk fences and a note on a trapline between Hawke Bay and Deer Harbour. Other loose materials have been left out because they have no clear tie to Labrador matters: for example, an elaborate central hall and parlour-type residence with wings and enclosed courtyard resembling a Georgian style house; a plan for a flat-bottomed "gaming canoe" intended for hunting in England; a complex, labyrinth-like animal trap of unlikely use in Labrador; a list of items and their weights; and a brief list of trivia relating to Hudson's Bay.

12   Tucker's account follows other nineteenth century visitors' accounts of Labrador life in serving up a mix of some first-hand observation and many secondary

sources. Although his note on the value of a sealing berth is relevant, his description of sealing is of the boat-based type carried out in northern Newfoundland. Similarly, his descriptions of Innu and Inuit, with whom he had no contact, are based on secondary sources.

13 Three small islands lie just offshore and north of Cape Charles. Today's 1:50,000 National Topographic System map (3D/4E) names these the Seal Islands. Wall Island is the easternmost and closest to the mainland cape, Tilcey Island is in the middle, and Fox Island is the westernmost. In Cartwright's time "Seal Island" was probably today's Wall Island, while Tilcey Island was called White-Fox Island. In *Additions* White-Fox Island is referred to as Cabareta Island. Seal Island and White-Fox/Cabareta Island lie parallel to one another for a short portion of their length where seal nets could be strung across.

14 Nicholas Darby, in a late eighteenth-century petition to repossess confiscated goods and to justify having illegally hired Canadien sealers at his Cape Charles post, wrote that "when the French were possessed of that Coast these men Wintered on that Coast a great many years" (CO 194/18:82v). Further evidence of early trapping comes from Cartwright, who in 1770 recorded that along the north shore of the St Charles River, "we found two old furriers' tilts, and snow death-falls; which appeared to be of Canadian construction" (C.18.10.70).

15 The opening of the Hudson's Bay Company at North West River in 1836 was intended to attract Innu away from the French posts along the Gulf coast. The HBC also built interior outlier posts at Fort Naskapi, Michikamau, and Winokapau that were deliberately positioned in the central region to attract the Innu, with their furs, who would normally be headed to the southern missions. HBC posts along the northern coast of Labrador were similarly intended to draw Inuit trade away from Moravian stations. The Innu, however, did not flock to the HBC posts as hoped, preferring to journey to the gulf where extended families could reassemble and where long-held ties with their Catholic parishes could be maintained. The HBC responded by encouraging an Oblate mission at North West River, but when it closed in 1893, about two hundred Innu shifted their trade back to Sept-Iles. This resulted in the HBC encouraging Settler trappers to move into traditional Innu resource areas along the Churchill River, the Kenamu, and Sandwich Bay, which grew into the large-scale Settler furring industry of the mid-1900s (Zimmerly 1975; Mailhot 2001).

16 Early ethnographers Julius Lips (1936, 10) and John M. Cooper (1938) were unable to identify traditional use of a gravity-based trap pit among the Cree and Innu in the western part of the Labrador-Quebec peninsula, drawing the conclusion that they were European introductions. But clear descriptions of the deadfall and

other traps used by northern Algonquian and Athapaskan peoples are found in their reports as well as in Provencher (1953).

17 This description of a snow house resembles that printed in Cartwright's *Journal* (c.26.2.71). A nearly identical description was recorded by William Richardson, who visited George Cartwright in Labrador in 1771 (Richardson 1935). For comparison, other early descriptions of Inuit snow houses from the Arctic were recorded by Parry (1824) and Lyon (1825), while Jolliet recorded early (1694) descriptions of Labrador Inuit sod houses (Delanglez 1948).

TRANSCRIPTIONS

1 This entry is missing from Cartwright's table of contents but occurs in the manuscript.
2 Entry incorrectly paginated: should be on page 60.
3 This is the final entry in *Additions*, which Cartwright did not list in his table of contents.
4 Refers to a bird taken from a wild nest as a chick and raised in captivity.
5 A young bird that has reached full plumage.
6 Cartwright did not fully index the plan and the few numbers are too faint to transcribe. The original may have been colour coded.
7 A corner hole.
8 A collection of twenty-four or sometimes twenty-five sheets of paper of the same size.
9 A thin nail of the same thickness but tapering in width and with a slight projection at the top of one side instead of a head.
10 A powder made of the dried and crushed carapace of the Spanish fly or common blister beetle (*Cantharia vesicatoria*), which causes extreme skin irritation and blistering.
11 Yellow arsenic sulphate, highly toxic, used as a paint pigment and fly poison.
12 Savin is a type of juniper.
13 A name for European heather; Cartwright is probably referring to lichens or low bush varieties in Labrador.
14 A metallic element.
15 The stiffeners in corsets. Here it may refer to a device that holds furs to their shape.
16 William Paulson is listed in the Hudson's Bay Company Archives (Winnipeg) as having been at "Slude Fort," which was the alternate name for the Eastmain post on the Eastmain River. He served there as surgeon from 1783 to 1788, and also as

surgeon and master between 1785 and 1788. He returned to England in 1788 to his home parish of Mansfield in Nottinghamshire. Cartwright lived in Mansfield in his final years and undoubtedly met Paulson there.

17   Presumably Sir Alexander Mackenzie (1764–1820), *Voyages from Montreal* (London 1801).

18   Probably refers to a tobacco powder.

19   Barming is barley processing for beer brewing.

20   No sketch appears on page 108.

21   An opium based mixture; laudanum.

22   Unguent of mercury, nitric acid, and a paste base such as lard.

23   William Cartwright (b.?–d. 1781), a former high sheriff of Nottingham, was George Cartwright's father.

24   *Carcajou* is the Algonquin name for wolverine.

25   The name of Cartwright's vessel.

26   The inscription reads:

> In memory of George Cartwright, Captain of His Majesty's 37th Regiment of Foot. Second son of William Cartwright Esq. of Marnhem Hall in Nottinghamshire who in March 1770, made a settlement on the coast of Labrador, where he remained for sixteen years. He died at Mansfield in Nottinghamshire the 19th of February 1819. [The date of death given here may be incorrect as George was buried on 24 May 1819.]
>
> Also of John Cartwright, Lieutenant of the Guernsey, five years Surrogate of Newfoundland, and afterwards Major of the Nottinghamshire Militia. He died on the 23rd of September 1824.
>
> To these distinguished brothers, who in zealously protecting and befriending, paved the way for the introduction of Christianity to the Natives in these benighted regions. This memorial is affectionately inscribed by their niece Frances Dorothy Cartwright.

27   Twelve pence equal one shilling, and twenty shillings are one pound; 1/6 refers to one shilling and six pence. The term "pence" is shown as "d."

❊{ BIBLIOGRAPHY }❊

ABBREVIATIONS

CO   Colonial Office Archives
DCB   *Dictionary of Canadian Biography* (www.biographi.ca/EN/index.html)
DRO   Dorset Record Office
HBC   Hudson's Bay Company Archives
JBRP Joseph Banks Research Project
LAC   Library and Archives Canada
PANL Provincial Archives of Newfoundland and Labrador
PC   Privy Council

ARCHIVAL

Dorset Record Office
Lester and Garland Families Archive. D/LEG/x8; http://www.a2a.org.uk/

Hudson Bay Company Archives
Nachvak Post, 1873. B138

Joseph Banks Research Project
Correspondence from George Cartwright; http://www.a2a.org.uk/

Library and Archives Canada
Champlain, Samuel de. C-113066. Can be viewed at http://www.civilization.ca/vmnf/popul/coureurs/deer.htm
Colonial Office Archives. CO 191/31; 194/18, 19, 27, 31, 32, 35, 62. Series 194 can be viewed at http://www.swgc.mun.ca/nfld_history/CO194/
Dartmouth fonds. MG23, A1, vol. 2, reel H-993, 2339–484.
Lower Canada Land Papers RG1, Series L3L, vol. 28. Letter from Cartwright to E. Nepean, 4 January 1787, 15057–67.

MG1, Series B. France: Archive des Colonies, Lettres envoyées, vol. 99, 1754.
Nain Diary. Moravian Brethren fonds. Archival number R10993-0-9-E.
William Bart. Musgrave fonds.

Provincial Archives of Newfoundland and Labrador
Jerrett, E.V. Diary of a Labrador Trapper. 2 December 1926–1 June 1927, MG 317.

PUBLISHED AND ELECTRONIC REFERENCES

Anick, N. 1976. *The Fur Trade in Eastern Canada until 1870*. Manuscript Report no. 207, vol. 2. National Historic Parks and Sites Branch, Parks Canada.

Armitage, P. 1991. *The Innu (The Montagnais-Naskapi)*. New York: Chelsea House.

Armitage, P., and M. Stopp. 2003. "Labrador Innu Land Use in Relation to the Proposed Trans-Labrador Highway, Cartwright Junction to Happy Valley-Goose Bay, and Assessment of Highway Effects on Innu Land Use." At http://www.env.gov.nl.ca/env/env/EA%202001/pdf%20files/TransLabHighway%20-%20Phase%20III/Innu%20Land%20Use%20Component%20Study/innulandusereport.pdf.

Atwood, M., and C. Pachter. 1997. *The Journals of Susanna Moodie*. Toronto: Macfarlane Walter & Ross.

Barkham, S. de L. 1976. "Two Documents Written in Labrador, 1572 and 1577." *Canadian Historical Review* 57, no. 2: 235–8.

– 1978. "The Basques: Filling a Gap in Our History between Jacques Cartier and Champlain." *Canadian Geographical Journal* 96, no. 1: 8–19.

– 1980. "A Note on the Strait of Belle Isle during the Period of Basque Contact with Indians and Inuit." *Études Inuits Studies* 4, nos. 1–2: 51–8.

– 1984. "The Basque Whaling Industry in Labrador, 1536–1613." *Arctic* 37, no. 4: 515–19.

Biggar, H.P. [1901] 1972. *The Early Trading Companies of New France: A Contribution to the History of Commerce and Discovery in North America*. Reprint, New Jersey: A.M. Kelley.

Borlase, T. 1993. *The Labrador Inuit*. Happy Valley-Goose Bay: Labrador Integrated School Board.

– 1994. *The Labrador Settlers, Métis and Kablunângajuit*. Happy Valley-Goose Bay: Labrador Integrated School Board.

Braun, H. 1962. *Old English Houses*. London: Faber and Faber.

Buckle, F. 2003. *Labrador Diary, 1915–1925: The Gordon Journals*. Cartwright, Labrador: Anglican Parish of Cartwright.

Butler, Mr. 1878. *The Labrador Mission*. Montreal: Witness Printing House.

Candow, J.E. 1989. *Of Men and Seals: A History of the Newfoundland Seal Hunt*. Hull, Quebec: Canadian Parks Service.

Cartwright, F.D. [1826] 1969. *The Life and Correspondence of Major Cartwright*. 2 vol. London: Henry Colburn. Reprint, New York: Augustus M. Kelley.

Cartwright, G. [1792] 1993. *A Journal of Transactions and Events during a Residence of Nearly Sixteen Years on the Coast of Labrador*. Facsimile edition, Canadian Institute for Historic Microreproductions.

Cartwright, Mrs G. 1909. "Some Cartwright Records." *Transactions of the Thoroton Society, Nottingham*, 111–41. Also www.nottshistory.org.uk/articles/.

Cell, G. 1969. *English Enterprise in Newfoundland, 1577–1660*. Toronto: University of Toronto Press.

Charest, P. 1998. "Les Inuits du Labrador Canadien au milieu du siècle derniers et leurs descendants de la Basse-Côte Nord." *Etudes/Inuit/Studies* 22, no. 1: 5–35.

Clermont, N. 1980. "Les Inuits du Labrador méridional avant Cartwright." *Études/Inuit/Studies* 4, nos. 1–2: 147–66.

Cochrane, W. 1966. "A Trapper's Tilt." *The Beaver* 46, no. 1: 16–17.

Cockerill, A.W. 2004. "The Trappers of Labrador." *Material History Review* 60: 86–90.

Cook, R. 1995. *The Voyages of Jacques Cartier*. Toronto: University of Toronto Press.

Cooper, J.M. 1938. *Snares, Deadfalls and Other Traps of the Northern Algonquians and Northern Athapaskans*. Washington, D.C.: Catholic University of America.

Crowhurst, R.P. 1969. "The Labrador Question and the Society of Merchant Venturers, Bristol, 1763." *Canadian Historical Review* 50 (4 December): 394–405.

Davies, W.H.A. 1843. "Notes on Esquimaux Bay and the Surrounding Country." *Transactions of the Literary and Historical Society of Quebec* 4, part 1, 70–94.

Dawson, W.R. 1958. *The Banks Letters*. London: Trustees of the British Museum.

De Boileau, L. 1861. *Recollections of a Labrador Life*. London: Saunders, Otley & Co.

Deetz, J. and P.S. Deetz. 2000. *The Times of Their Lives: Life, Love, and Death in Plymouth Colony*. W.H. Freeman: New York. Also http://etext.virginia.edu/users/deetz/Plymouth/framing.html.

Delanglez, S.J. 1948. *Life and Voyages of Louis Jolliet (1645–1700)*. Chicago: Institute of Jesuit History.

DeVolpi, C.P. 1972. *Newfoundland: A Pictorial Record*. Sherbrooke: Longman Canada.

Feild, Bishop E. 1849. "A Visit to Labrador." *Church in the Colonies*, no. 19. London: Society for the Propagation of the Gospel.

Fitzhugh, L.D. 1999. *The Labradorians: Voices from the Land of Cain*. St John's: Breakwater.

Glassie, H. 1975. *Folk Housing in Middle Virginia*. Knoxville: University of Tennessee Press.

Gosling, W.G. 1910. *Labrador: Its Discovery, Exploration, and Development*. London: A. Rivers.

Goudie, E. 1973. *Woman of Labrador*. (1979 edition edited by David Zimmerly.) Toronto: PMA Books.

Goudie, H. 1991. *Trails to Remember*. St John's: Jesperson Press.

Great Britain. Parliament. House of Commons, Committee on the State of the Trade to Newfoundland. 1785–1808. *First, Second and Third Reports of the Committee Appointed to Enquire into the State of Trade to Newfoundland, 1793*. First Series Reports, House of Commons, vol. 10. Miscell., 392–503.

Grenfell, W.T. 1969. *A Labrador Doctor*. London: Hodder & Stoughton.

Grey, W. 1858. *Sketches of Newfoundland and Labrador*. Ipswich: S.H. Cowell, Anastatic Press.

Hallock, G. 1861. *Harper's New Monthly Magazine* 22, no. 131: 577–99.

Harris, R.C., ed. 1987. *Historical Atlas of Canada*. Vol. 1. Toronto: University of Toronto Press.

Innis, H.A. 1930. *The Fur Trade in Canada: An Introduction to Canadian Economic History*. New Haven: Yale University Press.

Kaplan, S.A. 1980. "Neoeskimo Occupations of the Northern Labrador Coast." *Arctic* 33, no. 3: 646–58.

– 1983. "Economic and Social Change in Labrador Neo-Eskimo Culture." Ph.D. dissertation, Department of Anthropology, Bryn Mawr College, Bryn Mawr.

– 1985. "European Goods and Socio-Economic Change in Early Labrador Inuit Society." In *Cultures in Contact*, edited by W.W. Fitzhugh, 45–69. Washington: Smithsonian Institution Press.

Kennedy, J. 1985. "Northern Labrador: An Ethnohistorical Account." In *The White Arctic, Part 2: Labrador*, edited by R. Paine, 264–305. St John's: Institute for Social and Economic Research, Memorial University of Newfoundland.

Kennedy, J.C. 1995. *People of the Bays and Headlands: Anthropological History and the Fate of Communities in the Unknown Labrador*. Toronto: University of Toronto Press.

– 1996. *Labrador Village*. Prospect Heights, Ill.: Waveland Press.

Leacock, E.B., and N. Rothschild. 1994. *Labrador Winter: The Ethnographic Journals of William Duncan Strong, 1927–1928*. Washington: Smithsonian Institution Press.

Lips, J. 1936. *Trap Systems among the Montagnais-Naskapi Indians of Labrador Peninsula*. Stockholm: Statens Etnografiska Museum.

Lyon, G.F. 1825. *The Private Journal of G.F. Lyon, during the Recent Voyage of Discovery under Captain Parry*. London: John Murray.

Lysaght, A.M. 1971. *Joseph Banks in Newfoundland and Labrador, 1766: His Diary, Manuscripts and Collections*. Berkeley: University of California Press.

Mailhot, J. 2001. *The People of Sheshatshit*. (English reprint edition, originally published 1993.) St John's: Institute of Social and Economic Research, Memorial University.

Marshall, I. 1979. Inventory of the Cartwright Papers. On file, Centre for Newfoundland Studies, Memorial University of Newfoundland, St John's.

– 1996. *A History and Ethnography of the Beothuk*. Montreal: McGill-Queen's University Press.

Martijn, C. 1980. "La Présence Inuit sur la Côte-Nord du Golfe St Laurent à l'époque historique." *Études/Inuit/Studies* 4, nos. 1–2: 194–8.

McAleese, K. 1991. "The Archaeology of a Late Eighteenth Century Sealing Post in Southern Labrador: Captain George Cartwright's 'Stage Cove.'" Master's thesis, Department of Anthropology, Memorial University of Newfoundland.

Mellin, R. 2003. *Tilting: House Launching, Slide Hauling, Potato Trenching, and Other Tales from a Newfoundland Fishing Village*. New York: Princeton Architectural Press.

Moravian Mission Paper. 1962. "Voyages to Labrador in the Years 1765 and 1770." London: The Trust Society for the Furtherance of the Gospel. Microform 512, Centre for Newfoundland Studies, Memorial University of Newfoundland.

Niellon, F. 1996. "Du territoire autochtone au territoire partagé: Le Labrador, 1650–1830." In *Histoire de la Côte-Nord*, edited by P. Frenette. Collection Les Régions du Québec 9. Quebec: Les Presses de l'Université Laval.

Niellon, F., and M. Lamontagne 1982. *Recherche archéologique sur la Basse Côte-Nord: La fouille de 1982 aux postes de Brador et de L'Île Bois*. 2 vols. Report on file, Québec, Ministère des Affaires Culturelles.

Noble, L.L. 1861. *After Icebergs with a Painter: A Summer Voyage to Labrador and around Newfoundland*. New York: D. Appleton and Co.

Nottinghamshire. 2006. http://www.nottshistory.org.uk.

Ommer, R.E. 1991. *From Outpost to Outport: A Structural Analysis of the Jersey-Gaspé Cod Fishery, 1767–1886*. Montreal: McGill-Queen's University Press.

Parry, W.E. 1824. *Journal of a Second Voyage for the Discovery of a North-West Passage from the Atlantic to the Pacific*. London: John Murray.

Pasteen, T. n.d. http://www.innu.ca/biography/tshishennishpasteen.html.

Pastore, R.T. 1987. "Fishermen, Furriers and Beothuks: The Economy of Extinction." *Man in the Northeast* 33: 47–62.

– 1989. "The Collapse of the Beothuk World." *Acadiensis* 19, no. 1: 52–71.

Pearson, A.A. 1978. "John Hunter and the Woman from Labrador: The Story behind a Picture." *Annals of the Royal College of Surgeons of England* 60: 8–11.

Periodical Accounts Relating to the Mission of the Church of the United Brethren, Established among the Heathen 1790–1796. 1797. Vol. 1. London: Brethren Society for the Furtherance of the Gospel.

Prior, A.H. 1925. *Mansfield Parish Church.* Mansfield.

Privy Council (Great Britain) Judicial Committee. 1927. *In the Matter of the Boundary between the Dominion of Canada and the Colony of Newfoundland in the Labrador Peninsula, between the Dominion of Canada of the One Part and the Colony of Newfoundland of the Other.* 12 vols. London: W. Clowes & Sons.

Provencher, P. 1953. *I Live in the Woods.* Fredericton: Brunswick Press.

Prowse, D.W. [1895] 2002. *A History of Newfoundland.* London: Macmillan and Co. Reprint, St John's: Boulder Publications.

Rapport de l'Archiviste. 1922–23. "François Martel de Berhuoage, Brouague ou Brouage, Commandant au Labrador." *Rapport de l'Archiviste de la Province de Québec pour 1922–1923.* Québec: Archives de Québec.

– 1943–44. *Rapport de l'Archiviste de la Province de Québec pour 1943–1944.* Québec: Archives de Québec, 171–206.

Richardson, S.C. 1935. "Journal of William Richardson Who Visited Labrador." *Canadian Historical Review* 16: 54–61.

Robertson, S. 1843. "Notes on the Coast of Labrador." *Transactions of the Literary and Historical Society of Quebec* 4, part 1, 27–56.

Rollman, H. 2002. *Labrador through Moravian Eyes: 250 Years of Art, Photographs and Records.* St John's: Special Celebrations Corporation of Newfoundland and Labrador, Government of Newfoundland and Labrador.

Rompkey, R. 1991. *Grenfell of Labrador: A Biography.* Toronto: University of Toronto Press.

– 1987. "The Canadian Imprint of George Cartwright's 'Labrador': A Bibliographical Ghost." *Canadian Poetry* 21. http://www.canadianpoetry.ca.

Rothney, G.O. 1934. "The Case of Bayne and Brymner: An Incident in the Early History of Labrador." *Canadian Historical Review* 15: 264–75.

Roy, P.-G. 1940. *Inventaire de pièces sur la côte de Labrador.* Québec: Archives de la Province de Québec.

– 1942. *Inventaire de pièces sur la côte de Labrador.* Québec: Archives de la Province de Québec.

Ryan, S. 1994. *The Ice Hunters: A History of Newfoundland Sealing to 1914.* St John's: Breakwater Press.

Steffler, J. 1992. *The Afterlife of George Cartwright.* Toronto: McClelland & Stewart.

Stopp, M.P. 1995. "Long Term Coastal Occupancy in Southern Labrador." On file, Provincial Archaeology Office, St John's.

– 2001. "Lettuce and Labrador." *The Beaver* 81, no. 2: 27–30.

– 2002. "Reconsidering Inuit Presence in Southern Labrador." *Études/Inuit/Studies* 26, no. 2: 71–106.

– 2004. "Final Report of the 2004 Excavation at Lodge 1 (FbAx–04)." On file, Provincial Archaeology Office, St John's.

Story, G. 1980. "'Old Labrador': George Cartwright, 1738–1819." Lecture delivered to the Newfoundland Historical Society, Thursday, 9 October 1980. On file, Centre for Newfoundland Studies, Memorial University of Newfoundland, St John's.

Tanner, V. 1947. *Outlines of the Geography, Life and Customs of Newfoundland.* Vol. 2, *Labrador.* Cambridge: University Press.

Taylor, J.G. 1977. "Traditional Land Use and Occupancy by the Labrador Inuit." In *Our Footprints Are Everywhere*, edited by C. Brice-Bennett, 49–58. Nain: Labrador Inuit Association.

– 1983. "The Two Worlds of Mikak, Part 1." *The Beaver* 314, no. 3: 4–13.

– 1984. "The Two Worlds of Mikak, Part 2." *The Beaver* 314, no. 4: 18–25.

Taylor, V.R. 1985. *The Early Atlantic Salmon Fishery in Newfoundland and Labrador.* Canadian Special Publication of Fisheries and Aquatic Sciences 76. Ottawa: Department of Fisheries and Oceans.

Them Days. www.themdays.com

Thornton, P. 1977. "The Demographic and Mercantile Bases of Initial Permanent Settlement in the Strait of Belle Isle." In *The Peopling of Newfoundland: Essays in Historical Geography*, edited by J. Mannion, 152–233. St John's: Institute of Social and Economic Research.

– 1990. "The Transition from the Migratory to the Resident Fishery in the Strait of Belle Isle." *Acadiensis* 19, no. 2: 92–120.

Tooker, E. 1967. *An Ethnography of the Huron Indians, 1615–1649.* Midland: Huronia Development Council and the Department of Education.

Townsend, C.W. [1911] 2003. *Captain Cartwright and His Labrador Journal.* Boston: Dana Estes & Co. Reprint, St John's: DRC Publishing.

Trudel, F. 1978. "The Inuit of Southern Labrador and the Development of French Sedentary Fisheries (1700–1766)." Paper no. 40, National Museum of Man Mercury Series. Canadian Ethnology Service, Ottawa.

Tucker, E. 1839. *Five Months in Labrador and Newfoundland, during the Summer of 1838.* Concord, N.H.: I.S. Boyd and W. White.

Turgeon, L.W. 1994. "Vers une chronologie des occupations Basques du Saint-Laurent du XVIe au XVIIIe siècle – Un retour à l'histoire." *Recherches amérindiennes au Québec* 24, no. 3: 3–15.

Turgeon, L.W. Fitzgerald, and R. Auger. 1992. "Les objets des échanges entre Français et Amérindiens au XVIe siècle." *Recherches amérindiennes au Québec* 22, nos. 2–3: 152–67.

Vera, J.A.H., J.J.B. Calvo, J.N. Marcen, I.Z. Igartua. 1986. "Basque Expedition to Labrador 1985." In *Archaeology in Newfoundland & Labrador 1985*, edited by J. Sproull Thomson and J.C. Thomson, 81–98. Historic Resources Division, Government of Newfoundland and Labrador.

Whiffen, M. 1960. *The Eighteenth-Century Houses of Williamsburg.* Williamsburg: Colonial Williamsburg.

White, J. 1926. *Forts and Trading Posts in the Labrador Peninsula and Adjoining Territory.* Ottawa: F.A. Acland.

Whiteley, W.H. 1969. "Governor Hugh Palliser and the Newfoundland and Labrador Fishery, 1764–1768." *Canadian Historical Review* 50, no. 2 (June): 141–63.

– 1977. "Newfoundland, Quebec, and the Labrador Merchants, 1783–1809." *Newfoundland Quarterly* 73, no. 4 (Dec.): 17–26.

Zimmerly, D. 1975. *Cain's Land Revisited: Culture Change in Central Labrador, 1775–1972.* St John's: Institute of Social and Economic Research (ISER).